CONVERTING
AUTO ENGINES
FOR EXPERIMENTAL
AIRCRAFT

4th. Edition

By RICHARD FINCH, S.A.E.

MEMBER, SOCIETY OF AUTOMOTIVE ENGINEERS, CERTIFIED AIRCRAFT WELDER.
COMMERCIAL PILOT, MULTI-ENGINE, FLIGHT INSTRUCTOR, CERTIFIED AUTO MECHANIC
EAA TECHNICAL COUNSELOR, EAA FLIGHT ADVISOR

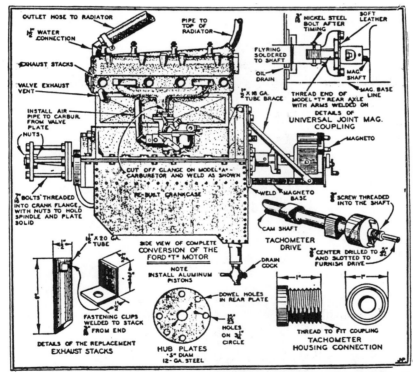

The drawing above is more than 60 years old! It is from the 1933 "Flying Manual"
Fawcett Publications

PUBLISHED BY FINCH BOOKS
PUBLISHER: RICHARD FINCH
TYPOGRAPHY: GAYLE FINCH
PHOTOS: RICHARD FINCH (EXCEPT AS NOTED)
COVER: AS NOTED

ISBN 0-9661457-1-2
COPYRIGHT 1998, FINCH BOOKS

LIBRARY OF CONGRESS, CATALOG NUMBER TX-1-518--565
PRINTED IN USA   FOURTH PRINTING

FINCH BOOKS
are published by
the Richard Finch Publishing Company
Santa Barbara, California
USA

This book is a fully revised and upgraded edition of
"Converting Auto Engines for Experimental Aircraft"
by Richard Finch, originally published in Titusville, Florida in 1985

This First Printing of the 4th. edition
is printed in 1997

© 1998 Richard Finch

Library of Congress Catalog-in-Publication

Finch, Richard, 1935
   Converting Auto Engines for Experimental Aircraft /
   Richard Finch - Revised 4th edition
        p. cm.
   Includes Index

ISBN-0-9661457-1-2

Book design and production by Richard Finch
Interior photos by Richard Finch, except as noted
Photo half-tones by Cecil Hatfield - 4 STAR PRINTING, Thousand Oaks, California
CAD drawings by Bud Rinker - Santa Barbara, California

CONVERTING AUTO ENGINES
for EXPERIMENTAL AIRCRAFT   4th. Edition

By Richard F. Finch, S. A. E.

ACKNOWLEDGMENTS: We are pleased to acknowledge the companies and individuals who have contributed pictures and facts to make this 4th Edition as comprehensive and helpful to the experimental aircraft builder as possible: Jess Meyers, Belted Air Power; Reiner Hoffman, Stratus, Inc.; Ken Duttweiler Automotive; Jeff Ackland, Legend; Ryan Falconer, Falconer Racing Engines; Tracy Crook; Donald Rivera, Airflow Performance; C. Hall Jones; Electromotive; Howell Engine Developments; Elwyn Johnson, Northwest Aero Products; Olsen Technologies; Ivo Zdarsky, IVOprop; Buick Motor Division & Oldsmobile Division, General Motors Corp.; and to my wife, Gayle Finch, for enduring the long hours it takes to word process a book like this.   *Richard Finch*

CONVERTING AUTO ENGINES FOR EXPERIMENTAL AIRCRAFT

This real-life Model A Ford engine is exactly like the one pictured on the first page of this book. See how simple auto conversion used to be. This book tells you how to make easy, simple auto engine conversions today.

PREFACE

Several weeks ago, a customer called me from the other side of the USA and asked me to help him make an educated, cost effective decision about the best way to go in selecting an engine for his almost completed BD-4 homebuilt airplane. He said his choices were between an 180 horsepower certified Lycoming engine for $20,000, or an auto engine, and he did not know what a ready-to-fly auto engine would cost.

His questions were very easy for me to answer because he knew how much horsepower he could afford and how much horsepower his BD-4 needed. So, here is what I told him:

CERTIFIED AIRCRAFT ENGINES

"If you buy a certified 180 horsepower 4 cylinder aircraft engine, you can expect a series of FAA mandated Airworthiness Directives that would force you to disassemble your $20,000 aircraft engine and inspect it for defects such as brittle rod bolts, brittle piston wrist pins, defective oil pumps, and quite likely something new and worse every 4 to 6 months during the life of the engine. You will be expected to pay for all the repairs because the factory will not warranty the engines it sells.

"There will be many more problems to solve in the projected 2,000 hour life of your certified aircraft engine, such as bad magnetos, bad fuel injector parts, bad wiring harnesses, and bad rocker arms, push rods and valves, and your $20,000 certified 180 horsepower aircraft engine will be hard to start when it is cold and when it is hot. It will always run rough, even rougher than your daddy's farm tractor engine does. And if you want a constant speed prop for the engine, the prop will weigh 80 pounds and will cost you $8,000 or more. Total cost could be $35,000 or more.

"But you can feel that you have invested in one of those "Dependable, Certified Aircraft Engines".

V-6 AUTO ENGINES

I then told him that if he decided on a factory new V-6 Auto Engine, Ford, Chevrolet or Buick, he would pay $1,500 or less for the long block engine, $2,500 for the prop belt drive unit, $1,900 for a 3-blade, carbon fiber, electric in-flight adjustable prop, another $1,000 for a starter, intake manifold, carburetor, alternator, exhaust system, and $500 for an engine mount. The radiator would cost $400; for a total investment, firewall forward, of $7,800 !

And his V-6, 200 horsepower auto engine would not have any Airworthiness Directives to comply with, it would run exactly as smooth as the engine in a new car or pickup truck, and, according to the past history of more than 60 years of flying auto engines in airplanes, the V-6 engine will last over 1,000 hours, possibly 2,000 hours.

I also told him that if he should decide at a later date that his BD-4 would be nice to fly if it had 250 horsepower, or even 300 horsepower, his auto engine could be hopped up to produce a dependable 250 to 300 horsepower.

(Ten years and over 10 million miles of racing the 3.8 to 4.1 Buick V-6 engine in the Indy Lights open wheel race car has proven the Buick V-6 will produce 425 horsepower with **zero** failures!)

Of course, when you hop up a V-6 engine with forged pistons, forged crankshaft, titanium valves, and other hop-up parts, the costs go up, as does the fuel consumption rate.

So, in this new 4th edition, I have tried to tell you exactly how to take advantage of this hot new trend in aircraft power, the auto engine! If you carefully study each chapter of this book, and if you can build your own experimental airplane, you should be able to duplicate the successful auto engine conversions pictured and explained in this book.

Happy Flying to You!

CONVERTING AUTO ENGINES FOR EXPERIMENTAL AIRCRAFT

You will likely recognize this aircraft engine as a Rolls Royce Merlin V-12, ready to install in Air Race Pilot and Owner Micky Rupp's P-51 Mustang. This engine is water cooled and has a PRSU gearbox to allow the engine to turn faster than the propeller, for more power and more top speed.

This V-12 appearing engine in a P-51 Mustang Replica is in fact a 454 cubic inch Chevrolet V-8, but it looks and sounds just like a real V-12 Rolls Royce Merlin. The P-51 Replica is a Stewart S-51 80% scale version.

Dependability Considerations

◆◆◆◆◆◆◆◆◆◆◆◆◆◆◆◆◆◆◆◆◆◆◆◆◆◆◆◆◆◆◆

The airplane to the left and below is the Oldsmobile V-8 powered PROWLER designed and built by George Morse of Redding, California, in 1975. It has flown dependably for over 20 years, and now has over 500 hours of time on its aluminum block, 264 cubic inch engine. The airplane above and to the right is the Rodeck (small block Chevrolet V-8) engine Jaguar that has also been flying dependably for over 10 years. Both airplanes have engines that are water cooled and both would perform dependably in a car that you could drive on the street or the highways. For more information about these two very dependable airplanes, check the various photos throughout this book and look up George Morse in Chapter 19.

Photo: George Morse, Prowler Aviation

A pair of Buick V-8 engines in flight over the Santa Paula, California orange groves. The Globe Swift, above and to the left, is piloted by Jess Meyers, of Belted Air Power, a 16+ year old company in Las Vegas, Nevada. The RV-6A to the right and below is piloted by Glenn Smith. Both airplanes are powered by 1963 215 cubic inch Buick V-8 engines, which are almost completely stock.
Photo: Helen Hutchings; photo plane pilot: Richard Finch

VALID ARGUMENTS

One of my older (aged) friends, who is a former Lear Jet pilot and highly experienced in airplanes, once commented to me that he had never exceeded 2,800 rpm on his Buick Park Avenue 4 door sedan. He had doubts that his Buick car would be dependable if he cruised it at 4,500 rpm and 120 miles per hour.

Another friend and co-worker of mine once told me that he had never, never in his entire life,

exceeded a posted speed limit, and he was 55 years old when he told me that! He also told me that he had never owned a car with a radio in it or with a heater in it! (I do need to tell you that he had lived all his life in Melbourne, Florida, where it seldom freezes.)

Many other people sincerely believe that auto engines are designed for low power, low rpm operation. That is not a correct assumption, and I will explain that in this chapter. But read about another objection to the subject of this book.

WORLD FAMOUS MAGAZINE

In as late as 1996, one of the foremost magazines in the world published a 10 page article titled "WHY AUTO ENGINES WILL NEVER FLY". The article gave 10 pages of examples of failures of automobile engines in airplanes. Suffice it to say that the article could and did not fully explain all the reasons for airplanes that crashed while being flown with auto engine power. One instance was reporting that the Nissan V-6 powered Pond

This 1935 model airplane is the Ford Arrow Sport, an airplane that used a stock 85 horsepower flathead Ford V-8 engine as its certified airplane engine.

Racer crashed at the Reno Air Races. No specific cause of the crash was reported in the article, but reports published elsewhere gave the cause of the crash as propeller blade failure, which had absolutely nothing to do with auto engines in the Pond Racer. The propeller blade that came loose was an airplane part, not an automotive part.

Another example of auto engine failure reported in the article concerned the ill-fated Thunder Engine program of the 1970 - 1980 era. The magazine article reported that Dick Mc Coon spent $10 million dollars and could never get his engines to perform dependably. The author of the article intimated that the cause of failure was the lack of auto engine dependability.

However, the article failed to mention that the same basic engine was flown very successfully by Jeff Ackland in his Legend Airplane. The Legend is featured in several chapters of this book.

The conclusion that an average airplane person should get from these two examples, the Thunder Engine and the Legend Engine, is that Jeff Ackland knows how to successfully convert an auto engine for use in experimental airplane, and that Thunder Engine could not or did not want to be successful at the same job.

SOME CAN, SOME CAN'T

The general approach to the subject matter in the 4th Edition of this book is to show that there are lots of auto engine conversion people who get the job done with little or no trouble.

Then the book will make mention of others who loudly proclaim that converting an auto engine for use in airplane is an impossibly difficult task.

I'll tell you right now that you should not listen to the people who say it can't be done. You should be polite to those detractors, but ignore most of what they say. You should listen to those who have made this movement a success. As I said in the first and second editions, "If you want to win (a race) copy the winner and ignore the loser".

NEGATIVE PEOPLE

Back to the subject of the magazine article: I could not let the article go as printed, so I wrote a letter to the editor, offering to write an opposing view article in a subsequent issue. The editor answered my letter quickly, but refused to acknowledge that the September-October 1996 article was negative!

And he never published my letter to the editor. His two letters to me asked, "If auto engines are dependable, why don't the auto manufacturers <u>try</u> to certify them for airplanes? There is an answer to that negative question...see the next paragraph.

AUTO MANUFACTURER CONVERSIONS

During the past 60 to 70 years, a significant number of automobile manufacturers have actually certified their engines for use in certified airplanes.

Most recently, the Japanese Auto manufacturer, Toyota, certified the Lexus V-8 engine for use in aircraft. The expected selling price of the Toyota Lexus V-8 is about $60,000 or more...think about the price. Several years ago, the German car maker, Porsche, certified an air cooled 6 cylinder derivative of the Porsche model 911 engine for use in certified aircraft. The selling price for this very quiet and smooth running engine was in the neighborhood of $50,000. Think about the price for a minute...records show that Porsche sold less than 100 engines, probably about 25, before they stopped production.

CERTIFIED AUTO ENGINES

In the early days of aviation, during the 1930's, several auto manufacturers in the United States of America certified auto engines for use in certified airplanes. Ford Motor Company certified the 85 horsepower flathead Ford V-8, Crosley certified the little 4 cylinder inline Crosley engine, and others installed their engines in airplanes and flew them as publicity and advertising promotions.

Hudson, Terraplane, Plymouth, Studebaker, and many others flew their engines in Curtiss Robin Airplanes, Fairchilds, and in special designs.

PACKARD AVIATION ENGINES

The Packard Motor Car Company of the United States actually manufactured all sizes of certified and military aircraft engines for over 40 years. The Liberty V-12 aircraft engine was manufactured by

Packard, as were thousands of Packard-Merlin V-12 engines during WW II. Packard merged with Studebaker in 1954, and then ceased to exist. A very wonderful hard cover book about the history of the Packard Motor Car Company details the many, many exquisite airplanes that were powered with Packard aviation engines. Look for this book in your public library.

GENERAL MOTORS AVIATION ENGINES

The Allison Division of General Motors built thousands of Allison V-12 aviation engines in the WW II years. These engines were used in P-38 twin-boom, twin engine airplanes and in P-51 Mustang airplanes, and were even used as engines to propel tanks. If you research the numbers of aviation engines built, you will likely find that the automobile manufacturers have built more airplane engines than the more common airplane engine builders, Lycoming, Continental, and Franklin have built.

WHY NOT TODAY?

The next question, and one that was asked by the editor of that magazine is, "Why don't Ford, Chevrolet and other auto engine manufacturers certify current auto engines for aircraft use today?"

The answer is in numbers (see page 12 for those numbers).

CONTROVERSY

The debate still · rages, maybe with even more anger than ever before. The subject: Are auto engines as dependable as aircraft engines? Here are the facts:

FACT ONE

As the recognized "curator" of auto engine conversions used in airplanes, I can truthfully tell you that there are no reports of an auto engine wearing out when correctly modified for use and flown in an airplane. That is, if the engine was new when it was installed in an airplane, or properly zero-time overhauled to like-new condition.

This fact includes the Pietenpol with Model A and Model T

10

Ford engines, and it includes long-time installations such as the Buick V-8 in Steve Wittman's Tailwind. But even if an auto engine would only last 1,000 hours in an airplane, it would still be less expensive to buy a brand new replacement auto engine at 1,000 hours flying time than to top overhaul an equal horsepower certified aircraft engine.

FACT TWO

There have been many experimenters who have successfully flown auto engine conversions with very little trouble. Some of these people are Bernie Pietenpol, who flew Fords and Chevrolet Corvairs; George Morse, who flew Buick V-8 and Chevrolet V-8 engines; Fred Geschwender, who flew all sizes of Fords as well as Chevrolet V-8 engines; Jess Meyers, who has flown Buick V-8 engines and Chevrolet V-6 engines. And there was Waldo Waterman, who flew the Chevrolet Corvair engine, Ole Falhan who flew the Plymouth engine, and certainly many more people who had no real

trouble getting automobile engines to fly successfully.

FACT THREE

Then there are quite a few people who incorrectly installed car engines in airplanes, had the inevitable failure, then shouted to the whole world that "It won't work!" Others in the past 70 years or more made obvious mistake after mistake in their specific approaches to flying auto engines, and were then very vocal in trying to convince the aviation world that they had overcome mountains of obstacles and spent thousands to millions of dollars in "engineering out the defects of auto engines".

So, why do so many find it relatively easy to fly an auto engine while still others find it impossible to do the same thing? There really ARE specific reasons, which will be explained in the next paragraph.

THE COMMON DENOMINATOR

The people who have trouble-free success in flying their

This water cooled V-8 engine, with racing type headers, is a famous OX-5 engine that powered many Curtiss Jenny airplanes in the 1930's.

Reiner Hoffman, owner of Stratus, Inc. in Seattle, Washington, has flown his Subaru Legacy auto engine powered Cessna 150 all over the USA   Photo: Stratus, Inc.

own designs of converted Auto Engines all have at least one thing in common about their success: THEY PROVIDE FOR EFFICIENT RADIATOR COOLING. This previous sentence is almost the number one secret of success that will be listed in this book. Take a close look at the good radiator design shown in Chapter 9. These designs WORK.

Those experimenters who have struggled with overheating problems, eventually come to realize that you must flow sufficient amounts of air through the radiator(s) and that you can not tolerate air traps in the plumbing.

The other common denominator that explains why many people have given up in disgust, is the common problem of over-modifying the engine to the point that it would not even perform in a car, much less in a boat or an airplane, has been high on the list of causes of failure.

COMMON QUESTIONS

People who are just becoming interested in flying auto engine power in experimental airplanes, usually have many of the same questions. First of all, they want to know how an auto engine in an airplane is cooled. Then they want to know how much heavier a 350 horsepower auto engine is compared to a 350 horsepower aircraft engine. Next, they want to know how often an auto engine in an airplane must be overhauled. And then, a significant question is whether an auto engine can be converted to do aerobatics.

AEROBATICS?

When asking about aerobatics, most neophyte experimental airplane people are referring to steep or even shallow turns, but some people are also asking about aileron rolls and about spins. They don't actually mean tail slides, hammerheads, and extended inverted flight.

CAST IRON CRANKS

The more informed builders often ask about cast iron (nodular) versus forged crankshafts, and about whether auto engine crankshafts will break. We all know that certified aircraft engine crankshafts break, and they often bend, with very expensive consequences if you have a prop strike.

In the following paragraphs, these common questions will be answered, and in the appropriate chapters in this book, the solutions to most of these and other questions will be presented. Although this 4th. edition of this book is more technical than the previous 3 editions, the solutions are sensible and practical.

AUTO ENGINE AEROBATICS

Certified aircraft engines, such as Lycomings, Continentals, and Franklins, must be specially equipped with dry-sump oiling systems in order to fly inverted for more than a couple of seconds. Auto engines can be equipped with the same inverted oil and fuel systems and can fly inverted also.

Steep turns, climbs, descents and positive G's maneuvers are aerobatic, but do not require an inverted oil system. An auto engine in an airplane will not have any oiling or fuel flow problems. 11

One of the high-time auto conversions is this Mazda Rotary powered Long Eze built by Ron Gowan of Roanoke, Texas. The Mazda engine had over 1,000 hours of flying time in 1996.

Ron Gowan's 13B Mazda rotary engine uses a Ross reduction gearbox and a 3 blade fixed pitch wood prop.

CAST IRON CRANKS

In the early days of Model A and Model T Ford engines used in aircraft, the engines occasionally experienced broken crankshafts because the big, slow-turning props were bolted directly to the crankshaft. Gyroscopic loads from the prop would weaken and break the crankshaft.

In the PSRU equipped, belt drive, chain drive, or gear drive units of today, there are no gyroscopic prop loads on the auto engine crankshaft. All the prop loads are taken in the PSRU, and the auto engine sees only torque loads, just as it sees when the car transmission is bolted to it. Do not be concerned about the materials of modern auto engine crankshafts. All auto engines of today are dependable.

THRUST LOADS

An old-time concern that auto engines could not handle the continuous thrust loads exerted by a propeller, were a reality in the 1920's and 1930's, when most auto engines used poured lead for the main crankshaft bearings. But since the late 1930's, auto engines have bi-metal bearings and the thrust bearings are very durable. Also, thrust loads on a crankshaft are a concern only if the engine does not utilize a PSRU.

When a PSRU is used, and in most cases they should be, the propeller thrust loads are absorbed in the ball and roller bearings of the PSRU and not in the engine crankshaft.

NUMBERS

As mentioned earlier in this chapter, the Big Three Automakers could not be concerned with selling auto engines to aircraft manufacturers because of the minuscule numbers of aircraft built each year.

In 1995, the entire General Aviation Manufacturers Association built only 515 single engine airplanes for the entire 'year. In contrast, Buick built more V-6 engines than that EVERY TEN MINUTES (!) for a total of 4,000 engines per day and over 8 million engines for the entire year!

Conversely, Lycoming advertised in 1978 that they had built "slightly over 230,000 engines in their first 50 years of operation. Chevrolet built more V-8 engines than that in one month in 1978. Why would auto engine manufacturers want a troublesome aircraft business?

TBO of AUTO ENGINES in AIRPLANES

The letters, TBO, stand for "Time Between Overhaul". Most certified aircraft engines are certified to fly for 1,200, 1,600, or 2,000 hours between overhauls. However, many certified aircraft engines require top overhauls long before the TBO time comes around. They

ELWYN JOHNSON C-172

Elwyn Johnson has been flying this C-172 for a number of years with a 4.3 liter Chevy V-6 marine engine installed. The engine has the balance shaft, and Elwyn also removed the stock cast-iron heads and replaced them with a pair of aluminum race car heads to save about 25 pounds of weight. The very dependable Chevy V-6 makes this C-172 airplane go very fast, in spite of the blunt nose cowl and the lack of wheel pants.

INDUSTRIAL USE

For many, many years, auto engines have been used to take the place of heavy, expensive, industrial engines in farm irrigation use, log splitting use, and especially in off-shore fishing boat use.

A 4.3 Liter Chevrolet V-6 engine powers this early Cessna 172 converted by Elwyn Johnson of Northwest Aero Products, Seattle, Washington. The C-172 will do 158 MPH with the Chevy, and that is faster than any 172 I know of.
Photo: Tom Watlanaja

Race car driver, A.J.. Foyt, Jr., drove this Oldsmobile Aurora powered race car to a new record for speed and endurance, equal to more than 50 Indy 500 mile races, stopping only for fuel and tires (and to sleep). NO certified piston aircraft engine is dependable or powerful enough to equal an auto engine in high speed endurance runs like this.

The Oldsmobile Aurora V-8 that powered the Olds race car pictured to the left. There is no certified piston driven aircraft engine that could last under these conditions. Auto engines are very durable.

do not make it to TBO.

More often than not, certified aircraft engines have cracked cylinders, valves that burn, rocker arms that wear out, and camshafts that wear out, and at very low times, such as 800 to 1,000 hours of operation. And, as an example, a 360 cubic inch certified aircraft engine cylinder, valves and pistons will cost $1,500 to $3,000, just for the parts, not including the labor to remove and replace them. (A new, complete V-6 auto engine only costs $1,500!) Certified aircraft engines seldom make to the TBO.

On the other hand, most late-model auto engines regularly exceed 100,000 miles, 150,000 miles, or even 200,000 miles of operation without an overhaul, 13

In coastal towns from Baja, California, to Anchorage, Alaska, fishing boats like this one regularly get over 2,000 hours of dependable service out of the Chevrolet and Buick V-6 and V-8 engines that power them. Dependability of the marine engine is extremely important for ocean going boats. An engine failure on the open sea can be as disastrous as an engine failure in an airplane.

even when they are used for high-speed highway travel, even to tow boats and travel trailers.

In a car or truck, 100,000 miles of travel at 70 miles per hour equals 1,428 hours of operation. 150,000 miles of operation at 75 miles per hour equals 2,000 hours of operation. Newer, late model auto engines seldom use oil at 150,000 miles of operation, and most auto engines will run past 100,000 miles without even so much as a tune up or even new spark plugs! Highway travel in the 1990's usually exceeds 75 mph.

PAST HISTORY, TBO

As mentioned elsewhere in this chapter, the record of dependable flight time by successful auto engine conversion people, shows that no auto engine has actually worn out, requiring an overhaul, in hundreds to thousands of hours of operation in airplanes.

Realistically, an auto engine can be expected to outlast a certified aircraft engine when used in an airplane. You may ask, why? That is because the auto engine manufacturers are constantly testing, and engineering, for long life and dependability, while certified aircraft engine manufacturers are doing little or nothing toward

14

improving their products.

No auto manufacturer in the world could afford the bad reputation of regular and consistent engine failures such as those that occur in airplanes powered by certified 4 and 6 cylinder air cooled aircraft engines. If, for instance, Ford auto engines were failing on a regular basis at 350 to 400 hours of operation (25,000 miles), people would refuse to buy Ford automobiles, and the world governments would force Ford to recall and repair 100% of their automobile engines. Auto engines are far more dependable than certified aircraft engines these days. In fact, the auto engine that is seldom used, and only to drive to the grocery store, is the engine that will give the least dependable service. Auto engines that spend 200,000 on the highway at high speed are usually trouble-free.

AUTO ENGINE FACTORY PROVING GROUNDS

Think back for a bit and you will remember the many magazine ads, newspaper ads, and recently, television ads by virtually all of the auto manufacturers that tout and advertise the DEPENDABILITY of their cars.

Ford Motor Company (Ford, Mercury, and Lincoln automobiles) has the Kingman, Arizona proving grounds and test track. General Motors (Chevrolet, Buick, Pontiac, Cadillac and Oldsmobile) has their Milford, Michigan and Phoenix Arizona test tracks. And all of the US auto manufacturers road test their new engine designs in the heat of the Southwestern Desert and in the cold of winter in places such as International Falls, Minnesota.

This author was recently employed by a small company here in California where we re-designed and destructive tested drive train components for most of the American and European auto manufacturers. We literally chained the rear axle to a fixed object and then burned rubber until we destroyed the gear-box, transaxle or the transfer case of the vehicles we were testing. Then we re-designed the gears, chain drive or axle shafts that failed, and then we tested again. And we tested on a test stand, then we gave the new vehicles back to the manufacturers so they could drive test them to destruction also. That is how all auto manufacturers prove their engines, gear boxes and suspensions. It is safe to say that your new 1999 to 2001 auto design has been tested hundreds of thousands of miles and many thousands of hours before it is put on the market.

AIRCRAFT ENGINE FACTORY TESTS

The Federal Aviation Administration allows all piston engine aircraft engines to be certified and sold with only 150 hours of testing of the prototype design. A new design aircraft engine can be certified on a test stand, without ever spending a single hour or even one .minute flying in the airplane it is certified for.

And those 150 hours are at 45% to 70% power. It is no wonder that modern certified aircraft engines are failing at an alarming rate. See pages 143 & 144 for specific figures on aircraft engine certification.

Choosing the BEST Auto Engine

◆◆◆◆◆◆◆◆◆◆◆◆◆◆◆◆◆◆◆◆◆◆◆◆◆◆◆◆◆◆◆◆◆◆◆

For airplanes that need 125 hp, this belt drive Honda Accord engine is a good choice. Avtech of Montana makes this conversion. It is very neat and compact.

If the choice of the best auto engine to fly in your airplane were easier to make, many, many more people would be flying auto engines than are being flown now. It is relatively easy to make a choice if several dozen specific experimental airplanes are successfully flying with a specific auto engine, but this is a rare situation. In many cases, you are making a breakthrough engine choice yourself.

GOOD EXAMPLES

Most of the pictures of auto engine powered airplanes in this book are airplanes that perform very well with the engines that are installed in them. To make an educated choice of the best engine to convert for your experimental airplane, read this chapter closely, and then GO FOR IT!

When the 1st edition of this book was written way back in 1984, the trend in auto engine conversion for aircraft was cost driven for sure. Ninety nine percent of the engines converted for aircraft use were used, wrecking yard engines. Perhaps a dozen airplanes were flying with purpose-built, high performance auto engines. As the technology matures, more and more builders are spending more and more money to be able to fly the best conversion possible.

In this chapter I will explain at least five ways to go in converting the particular engine you decide on. And these five ways all relate to the amount of dollars you have to spend. Here are five ways to go:

1. Low bucks, used engine, do it yourself: $1000
2. Factory new, firewall forward kit: $6,500
3. Race car prep, stock block & heads: $10,000
4. All-out after-market Race engine prep: $29,000
5. Custom engine, 12 cylinder & etc. $90,000

V-12 KR-2?

Ordinarily, the decision about which airframe comes before the decision about which engine to use in it for motive power.

Practicality is usually secondary to excitement when deciding on an experimental airplane. For some people, a heavy hauler that will take them to remote camping and hunting spots is the kind of airplane that excites them. Then, the choice of a big block V-8 with lots of power for short take-offs is the logical choice of engine.

If a slick little KR-2, two place that can be hauled home on a trailer and tuned up in your driveway excites you, then you probably need 100 to 150 horsepower at the most. That would narrow your engine choice down to a Honda or Toyota or GM 4 cylinder of about 125 to 175 cubic inches. You certainly would not want a 600 cubic inch V-12 engine for a KR-2.

GLASSAIR, LANCAIR, RV, ETC.

If a comprehensive survey should be taken, more than 50% of experimental aircraft builders would want a very fast, very roomy 2 passenger or 4 passenger, sexy-looking airplane. The balance of the builders want replica fighters, replica W W I bi-planes, ultralights, paragliders and other kinds of airborne craft.

For the builders and owners of the first 50%, Glassair, Lancair,

RV, etc., the choices of engines are plentiful. In this book you will see many examples of auto engine power in slick, fast airplanes.

OVERHEAD CAM ENGINES VS PUSHROD ENGINES

The trend in Japanese, German and several US auto engine designs is the dual and quad overhead cam engine with 4 or 5 valves per cylinder. While these designs produce a small percentage more horsepower than a standard pushrod, 2 valve per cylinder engine, they do so at the expense of higher R.P.M.'s, more parts required, more overall size, and in many cases, even more mass/weight.

For instance, the Cadillac Northstar V-8 engine is 50% wider, 20% taller, and 25% heavier than the 350 cubic inch Chevrolet V-8 pushrod engine. And, the Chevrolet V-8 produces 15% more power than the Northstar V-8 does.

The Buick 3800 V-6 engine is 25% smaller overall, 15% lighter, and produces 10% more power than a 4 cam all-aluminum engine. And the Buick V-6 is all cast iron. A four valve engine has twice as many cam and valve bearings, three times as many sprockets, pulleys, cam shafts and drive chain lengths, twice the valves, guides, lifters, keepers and 1.5 times the oil pump capacity as a 2 valve per cylinder pushrod engine of equal power. You should weigh these facts when you look for a suitable engine to convert for your airplane.

ALUMINUM VS CAST IRON AUTO ENGINES

Aluminum engines are often heavier than equal horsepower cast iron engines. The reason for this is that aluminum, although half the weight of cast iron per cubic foot of material, is less than half the strength of cast iron, therefore the thickness of aluminum castings must be doubled to maintain the strength of the engine.

One specific example compares the 300 pound, 2.5 liter Subaru Legacy (all aluminum) engine with the 285 pound, 4.3 liter Chevrolet V-6 (all cast iron) engine. The cast iron Chevrolet V-6 weighs less and produces a lot more torque and horsepower. You need to remember that cubic inches and cubic liters equal specific amounts of horsepower in piston engines.. A 2.5 liter, 153 cubic inch engine can produce 150 to 175 horsepower, while a 4.3 liter, 262 cubic inch engine can produce 250 to 275 horsepower. The 262 cubic inch cast iron Chevrolet V-6 engine would be the best choice for powering airplanes that need 200 to 250 horsepower.

HOW IS HORSEPOWER MEASURED?

The measurement of one horsepower is the ability to lift 33,000 pounds one foot high in one minute. Years ago, they figured a single horse could do that much work. Equivalent figures are 550 pounds lifted one foot in one second. The electrical equivalent is 746 watts and the heat equivilent is 2545 BTU

It might seem that this converted Honda Goldwing motorcycle engine would make a good aircraft engine, but it is quite heavy compared to the Honda Accord engine in the previous picture.

In other words, a certified 200 horsepower aircraft engine can lift 33,000 pounds (x 200) or 6,600,000 pounds in one minute! That is a lot of power. In numbers that are easier to relate to, a 200 horsepower engine can lift 110,000 pounds one foot high in one second. That is equal to lifting 44 average sized autos one foot off the ground in one second or 2,640 automobiles one foot off the ground in one 60 second minute. That is still a lot of work!

HOW IS TORQUE MEASURED?

The measurement of one foot pound of torque is exactly as it is stated. If an engine produces 200 ft. lbs. of torque, it will lift a 200 pound weight placed one foot outboard of the center of the turning crankshaft, in an instant.

Picture an axle shaft turned by the engine crankshaft at a 1 to 1 ratio of rotational speed. At the end of the axle shaft is a brake drum that turns with the axle shaft. Inside the brake drum is a backing plate with brake shoes attached to the backing plate.

If the brake shoes are engaged just enough to lift and hold a 200 pound weight placed at the end of a 12 inch (one foot) bar attached to the backing plate, you could say that the engine is producing at least 200 foot pounds of torque. Any more weight, even 5 more pounds, would begin to stall the engine and slow it down. Maximum torque usually is obtained at 50% to 75% of maximum engine rpm.

A normal, carburated, modern auto engine of 350 cubic inch displacement will usually produce 300 foot pounds of torque at 3,500 rpm, and 350 horsepower at 5,800 rpm, with a red-line, never exceed speed of 6,000 rpm.

HORSEPOWER & TORQUE CURVES

When an engine is tested on a dynamometer, the data collection system attached to the dynamometer will usually produce torque and horsepower curves. As stated in the previous paragraph, peak torque occurs at a lower rpm than peak horsepower, and both torque and horsepower peaks occur at about 10% to 20% less than a safe maximum rpm.

One fact that appears to be a phenomenon is that regardless of the size and power output of a modern auto engine, the torque and horsepower curves nearly always cross each other at exactly 5,250 rpm!

Take a look at the graph on page 18 for a typical torque and horsepower chart with plotted curves. Notice the 5250 rpm!

METRIC VS US HORSEPOWER & TORQUE

In recent years, the industry trend toward changing over to the metric system has served to confuse many people who once judged an engine by the horsepower it could produce at full throttle. In recent times, the automotive writers have been expressing the power output of an engine in kilowatts rather than horses. So, here is an easy conversion system so you can better understand kW Vs hp:

When you are studying magazine articles about new designs of auto engines, refer to this page for the formulas to help you better understand how much horsepower and torque a certain engine can produce. The formulas will work for English and Metric figures.

FORMULA FOR HORSEPOWER
* One horsepower = .746 kilowatts
* One hundred kilowatts = 134 hp (horsepower)
* One kilowatt (kW) = 1.34 hp
* To convert kilowatts (kW) to horsepower (hp), divide the kW by .746
* To convert horsepower (hp) to kilowatts (kW), multiply the hp by .746.

FORMULA FOR TORQUE
* One foot pound of torque = 1.355818 Newton meters
* One Newton meter of torque = .7375 foot pounds
* To convert foot pounds of torque to Newton meters of torque, multiply foot pounds by 1.355818
  *To convert Newton meters of torque to foot pounds, divide NM by 1.355818

FORMULA FOR CUBIC INCHES
* Multiply liters by 61 to get cubic inches (")
* 4.3 liters (L) x 61 = 262.3 cubic "
* 2.2 L (liters) x 61 = 134.2 cubic "
* 350 cubic " divided by 61 = 5.7 L

FORMULA FOR CUBIC CC'S OR LITERS
* Divide cubic inces by .061 to get CC's or Liters
* 231 cubic inches divided by .061 = 3786.8 CC
* 231 divided by 61 = 3.78 Liters

Because of its compact shape, many people convert the 2 rotor Mazda, but it seldom produces more than 125 to 150 hp, and it uses lots of fuel compared to a similar horsepower piston engine.

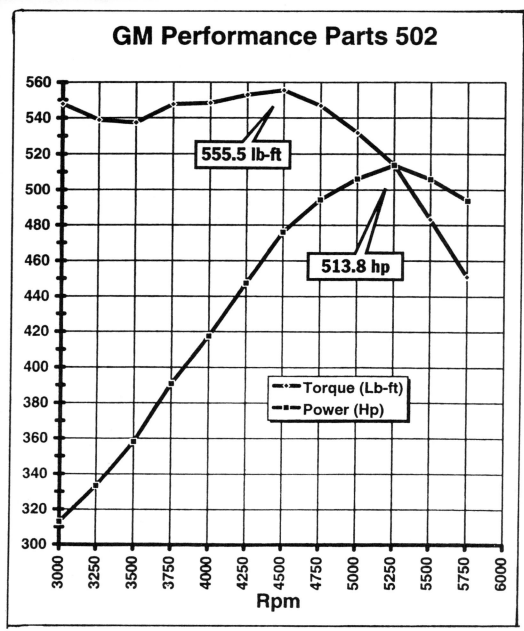

GM Performance Parts 502

555.5 lb-ft

513.8 hp

— Torque (Lb-ft)
— Power (Hp)

Rpm

This typical horsepower and torque chart for a General Motors 502 cubic inch V-8 big block engine points out the fact that torque and horsepower curves cross at 5250 rpm. The chart points out that torque is almost constant at all rpms while horsepower climbs with rpm until 5220 rpm, then it drops off. Also note a very important fact that this 514 hp engine is only developing 445 hp at 4250 rpm, and only 315 hp at 3000 rpm. Therefore, do not expect a direct drive engine to produce the maximum advertised horsepower at low rpm's, such as 2,750, etc.

WHAT IS TORQUE?

If you weigh 100 pounds and you stand on the 1 foot long lug wrench when you are changing a flat tire, you are exerting 100 foot pounds. That is 100 pounds applied 12 inches from the lug nut on a lever.

If you weigh 200 pounds and your lug wrench is 24 inches long, when you stand on the end of it, you are exerting 400 foot pounds of torque on the lug nut.

If the lug nut does not break loose, you are exerting torque at zero rpm. If you add a cheater bar to the lug wrench to make it 48 inches long, and you put all your 200 pounds on it, you have exerted 800 foot pounds of torque to the lug nut, and it will either come loose or break off!

Two hundred foot pounds of torque, but coupled with 4,000 rpm is what moves cars, airplanes and boats at high speed.

SPECIALTY ENGINES

The V-4 SCAT engine pictured at the top of this page is a custom-made engine, designed around the front and rear cylinders of a Chevrolet V-8 engine, with the center 4 cylinders left out. If you are interested in building the world's fastest Cassutt Racer or Formula One airplane, this engine might suit you.

LOW COST ENGINES

If your main goal is to get an airplane in the air, for pleasure flights on weekends, there is a simple solution. Take a look at the antique airplane at the bottom of this page. It flies very well with a 1962 Buick V-8 engine that usually can be purchased for $100 to $300 and then overhauled for another $150 to $200. The prop drive unit can even be made from parts scrounged from used 4-wheel-drive vehicle transfer cases. Other solutions for drive units are to build a belt drive unit, using an automobile rear axle shaft for the propeller shaft and then use bearings, belts and sprockets that you buy from a local bearing supply dealer. Costs for an entire belt drive unit can be as low a $750.

If you are looking for a lightweight, powerful engine for a sport or aerobatics airplane, this V-4 SCAT engine produces 250 hp, weighs 250 lbs. and is sold by SCAT Enterprises of Redondo Beach, CA.

For an absolutely low dollar conversion, check out this 215 cubic inch Buick V-8 powered airplane The Buick engine runs like a Buick car does. Cost of this engine is about $500.

FORD 3.8 - V-6

A relatively large number of 3.8 Ford V-6 engines have been flying successfully for over 20 years. If you decide on one of these engines for your aircraft, expect about 125 hp to 150 hp, and an installed weight of 425 lbs. Newer versions of this V-6 include a balance shaft to reduce vibrations.

A 3.8 liter Ford V-6 powers this very clean BD-4. A 3 blade Warp Drive prop looks good on it.

60 DEGREE V-6

The 60 degree V-6 engine design is very, very smooth in operation compared to a 90 degree V-6. A 60 degree V-6 engine is always smaller and more compact compared to a 90 degree V-6. The only negative thing about a 60 degree V-6 is that it is usually a smaller cubic inch engine. If you only need 175 to 200 horsepower and very light weight, consider a Chevrolet or other brand 60 degree V-6. This engine runs so smooth in stock operation that you can barely feel it running. And they come with aluminum cylinder heads for even lighter operation. The Chevrolet 60 degree V-6 is a good engine to substitute for a Lycoming O-320 or

an O-360.

It would make a good engine for a Lancair 235, 320, or a Thorp T-18.

CHEVROLET CORVAIR

Literally dozens of airplanes have flown for over 30 years with Chevrolet Corvair engine power. However, not all Corvairs are equal in power output. The Chevrolet factory built models from 80 horsepower in 1960, to 180 horsepower turbocharged in 1965 and 1966. The best Corvair engine is the 110 horsepower version that was manufactured from 1965 through 1969. In order to get the full power out of the Corvair engine, it should be geared down with a

Rinker gear box or a cog belt drive.

CORVAIR DEPENDABILITY

Over the past 30 to 40 years that Corvair Engines have been flying, there have been a few dependability problems. The most significant problem is the hardened valve seat coming loose in the cylinder head. This usually occurs after the engine has operated for at least 75,000 to 100,000 miles in a Corvair auto or van.

Don't overlook this smooth, smooth 60° Chevy V-6 engine, 3.1 Liter engine that weighs just 225 lbs. and produces 190 hp. Cost new: $1,400.

If you are looking for a gear drive Corvair engine, try this Bud Rinker design, now sold by Vertical Systems of Santa Barbara, CA. Corvair engines weigh about 200 lbs. and make very good airplane engines.

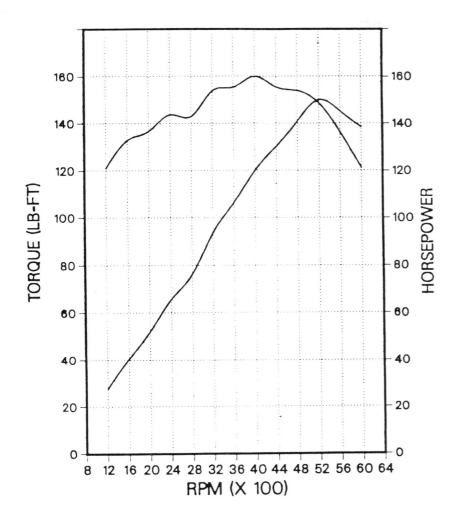

1988 LD2 2.3ℓ L4

RPM (X 100)

The horsepower and torque chart for the 1988 Oldsmobile Quad 4 shows that it produces 150 hp at 5,200 rpm, but only 85 hp at 3,000 rpm. You would need a PSRU with a 1.85 to 1 reduction ratio in order to take full advantage of the unusually high torque of this little 4 cylinder engine. Be sure to notice that the torque curve stays high from 800 rpm to 6,000 rpm. Re-grinding the camshaft(s) would not significantly improve the low rpm power output of this engine.

Graph, courtesy of Oldsmobile Engineering

$1,000 ENGINES

Quite a few airplanes have flown safely and successfully with engines that cost a total of less than $1,000 to purchase, overhaul and convert for aircraft use. Two good examples of low cost auto engines in airplanes are the Globe Swift and the RV-6A on page 8 of this book. The engines in those two airplanes are 215 cubic inch Buick V-8 engines. The cost to overhaul those engines was less than $300, which included new piston rings, new crankshaft bearings, a valve grind, and new gaskets.

In many other instances, Ford Escort 4 cylinder engines have been used, and the current cost for a low mileage Escort engine is less than $600. The belt drive systems were built from parts purchased locally at bearing supply houses.

$7,850 Engine Kits

Jess Meyers of Belted Air Power, Las Vegas, Nevada, has developed a firewall-forward Chevrolet V-6 engine kit that fits the RV-6A. Plans are to offer this same kit for other aircraft that use 160hp to 200 hp engines. Presently the kit sells for about $7,850.

An overlooked engine is the Toyota MR2 engine shown here. It is small, light and dependable, and it produces 1.5 hp per cubic inch. If you need 100 to150 hp in a very compact engine consider the Toyota MR-2 engine.

Oil leaks were once a problem with Corvair engines, but the fix described in "How To Keep Your Corvair Alive" effectively cures the leak problem. This calls for coating each push rod tube with ceramic paint to prevent the exhaust stack heat from hardening the O-rings.

The solution to the valve seat problem is to swedge the combustion chamber aluminum to hold the valve seat tighter. Another solution is to obtain .030" oversize valve seats, under cut the top half of the outer part of the seat, and then swedge the aluminum around the seat to make a permanent mechanical retention.

21

$10,000 ENGINES

In the photo at the bottom of this page you can see the author's Buick V-6 powered Grumman Traveler 4-place airplane. To duplicate this engine conversion with all new parts, the following costs would be incurred:

1. New 3.8 Buick long block engine $1,400
2. New intake, exhausts, starter, fuel injection, etc. $2,600
3. New gearbox or chain drive PSRU $3,000
4. New custom engine mount $500
5. New custom radiator, aluminum $700
6. Labor, extra parts $1,800

The above prices include balancing and blueprinting the engine to race car specifications. 220 hp expected.

$29,000 Engine

You can buy custom made racing engines in kits for $10,000 to $15,000. Then you need to add electronic fuel injection and a Geschwender PSRU for $8,000, and a few other custom items, and you have 500 horsepower V-8 to power your airplane. The Jeff Ackland Legend pictured in this book is a good example of such an engine.

$90,000 ENGINE

The Ryan Falconer V-12 engine pictured in several chapters of this book, is an example of the ultimate in non-aircraft, auto engine derived power for sport planes.

For those people who want to re-engine a good airplane such as the Piper Lance or the Piper Malibu, or to build an airplane to compete at the Reno Air Races, check out the Falconer V-12 engine. The applications that are possible with this engine are very exciting. You might even want to design your own 6 or 8 passenger airplane, using the V-12 as a power plant.

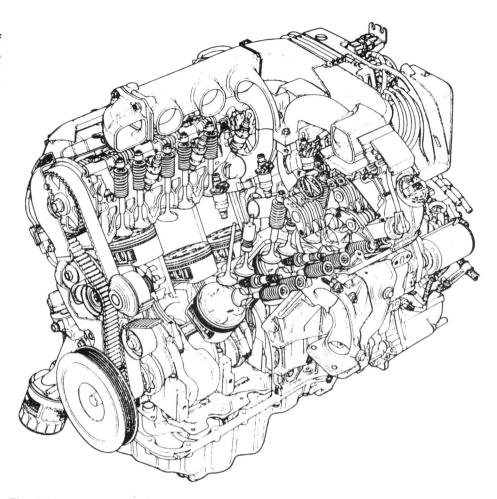

The 1998 Honda 60° V-6 engine is very smooth in operation, and it produces good power from its 3.0 Liter, 183 cubic inch displacement. It would make a good replacement for a 200 hp Lycoming or Continental aircraft engine. But weigh it first.

The author of this book has converted this 3.8 Liter, 220 hp Buick V-6 engine, connected to a prototype gearbox PSRU, also designed by the author. The conversion is installed in a Grumman 4 place airplane, licensed as an experimental. Cost of the Buick V-6: $1,800.

Once thought an impossible conversion, this 4 passenger Velocity aircraft is powered by a 4.3 Liter Chevy V-6 and connected to the electric Ivo Prop by a Northwest Aero Products belt drive PSRU. Cost of the conversion: $10,000.

Multi-valves in an engine may seem exotic, and they are, but they often add weight, size, and many extra parts to an otherwise simple, dependable engine design. Another negative thing about multi-cam and multi-valve engines is that they usually equal horsepower at only very high rpm.

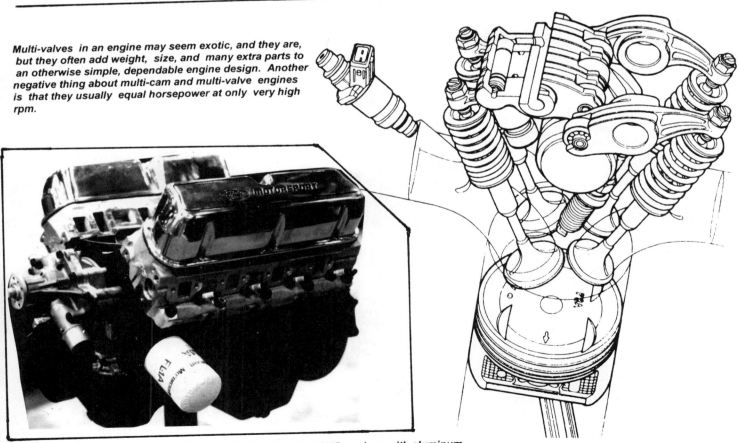

Your Ford Dealer will sell you this long block 302 GT-40 SVO engine, with aluminum heads, for $3,195. It weighs 430 lbs. and produces 320 hp.

23

LANCAIR ES-V-8

The airplane pictured at the top of this page has a Chevrolet Corvette engine in it that can be purchased for $3,495 according to one of the ads on page 26. The Lancair ES also uses a belt drive unit made by Northwest Aero Products, and that unit sells for $2,995. That is a lot cheaper than $60,000 for the certified Continental engine that the airplane was designed for. John Harlow of Tucson, Arizona is the builder.

CORVETTE LT-1 ENGINE

The photo in the middle of this page shows John Harlow's LT-1 engine and prop drive, a ready to install combination that sells for $6,490. That is a real bargain!

RODECK V-8

The engine in the lower picture is a small-block Chevrolet derived, 366 cubic inch all aluminum unit that is custom built by George Morse of Redding, California. It utilizes a 2-gear, PSRU that was designed by George. The complete package sells for under $20,000, and produces 350 horsepower.

The Rodeck V-8 block is used by many race car builders to obtain light weight and dependable engine performance. The crankshaft can be a forged unit or a billet steel unit, and the cylinder heads are usually by Dart or by GM

NOTE:

The top two photos are by John Harlow, Jr.

24

SUPER LANCAIR IV-P

This airplane won the runner-up (2nd place) award at the 1997 Sun-N-Fun air show for the 2nd best overall airplane at the show. And at Oshkosh 1997, the builders, Jim Rahm and his partner, of Daytona Beach, Florida, won the best engineered, most advanced airplane design. The airplane features a turbocharged V-8 and air conditioning, among other neat features.

LEGEND BIG BLOCK

Jeff Ackland of Olathe, Kansas, offered to sell duplicates of this 620 cubic inch Donovan/Brodix big block Chevy engine for $28,000. If there were such a thing as an equivalent certified aircraft engine (and there isn't!) it would cost upwards of $100,000, and the certified engine might break at any time. $28,000 is a real bargain.

FALCONER V-12

If I owned a 350 horsepower Piper Malibu airplane, I would pull out the troublesome certified aircraft engine and put one of these 600 cubic inch, 600 horsepower Falconer V-12 engines in it. Then I would go out and challenge other Malibu owners to races! This V-12 sounds gr-r-r-reat! and it really performs!

This engine as pictured below, is already equipped with electronic multi-part sequential fuel injection. It can also be purchased

with a P-51 Mustang type prop speed reduction gear box and even a 4 blade composite MT prop.

In addition to re-engineing production airplanes (the Aerostar twin would sound and fly great with this engine), special designs can be developed with this engine as the powerplant, up to 1,000 horsepower is possible from this engine, when supercharged or turbocharged. You

might want to call Falconer Racing Engines in Salinas, CA, for more information,

CHAPTER 2   Choosing the BEST Auto Engine

LISTED ON THIS PAGE ARE TYPICAL ADS FOUND IN MOST AUTOMOTIVE MAGAZINES. ALWAYS BUY ENGINES FROM REPUTABLE SUPPLIERS AND PREFERABLY NEAR YOU.

High Output 502ci (440 H.P.) #24502620   This new engine comes complete with roller cam, manifold, waterpump, balancer, flywheel, aluminum valve covers, steel crank and four-bolt main block.
• Also available is a High Output 454ci (425 H.P.) #24502618

New 502ci Premium Engine Kit #12371171   This 502 HP kit includes partial engine with roller cam, oval port aluminum heads and all items in the photo above.
• Also new is the 502/502 Base Engine, which includes partial engine and aluminum heads and related components #12371204

High Output 350 (355 H.P.) #24502609   This "new" ZZ4 engine features 10.1 compression, aluminum heads, aluminum manifold, water-pump, distributor, new piston and four-bolt main block.
• Also available is a Special Performance 350ci (300 H.P.) #12355345

See Chapter 19 for additional auto engine ads and sources of all types of auto engines that you can buy.

WEIGHTS of AUTO ENGINES

CHAPTER 3

◆◆◆◆◆◆◆◆◆◆◆◆◆◆◆◆◆◆◆◆◆◆◆◆◆◆◆◆◆◆◆◆◆

This 350 horsepower Chevy V-8 engine is many pounds LIGHTER than any 350 horsepower certified aircraft engine would be. George Morse of Redding, California has done a very professional job of converting the Rodeck engine to power his Jaguar Experimental airplane.

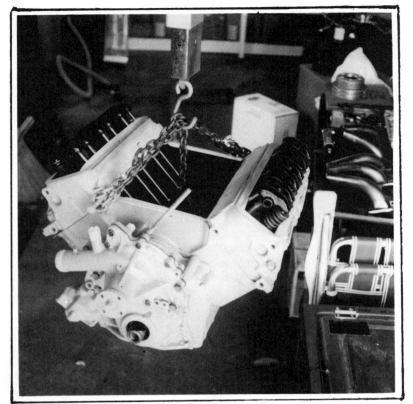

This 231 cubic inch all cast iron Buick V-6 LONG BLOCK engine is hanging from a scale, and the scale reads __260__ pounds. It is obvious that several accessories still must be installed to make the engine run. Ready to fly, this engine will weigh __365__ pounds, but its C.G. will be 10 inches aft of an equivalent horsepower aircraft engine.

WEIGHTS of AUTO ENGINES

The number one question that people ask about putting an auto engine in an airplane is, "But aren't auto engines a lot heavier than airplane engines?" The answer I give is that some auto engines are, in fact, a little heavier, but in many cases, the equivalent horsepower auto engine is substantially LIGHTER than the aircraft engine is.

LIGHT AUTO ENGINES

For instance, a 115 hp Subaru EA-81 auto engine is 20% lighter than a 100 hp Continental 0-200. A 350 hp Chevrolet V-8 is 30% lighter than a 350 hp Continental TIO-550 aircraft engine. A 160 hp PRV Renault V-6 is about the same weight as a 150 - 160 hp Lycoming 0-320 engine. And a 180 hp Chevrolet 60 degree V-6 is 5 to 10% lighter than a 180 hp Lycoming 0-360 engine. A 4.3 Liter V-6 is 10% heavier than a 160 hp Lycoming engine.

DEFINITIONS OF ENGINE EQUIPMENT

One of the rough things about converting auto engines for aircraft use is finding out what a ready-to-fly engine weighs. The following explanation should help you figure this out.

SHORT BLOCK: Includes the bare block, the crankshaft, the connecting rods, pistons, and wrist pins.

LONG BLOCK: Includes the short block, plus cam shaft, cylinder heads, valves, cam shaft, timing gears or timing chain, and usually the oil pump and water pump.

COMPLETE ENGINE: Includes the long block, plus oil pan, valve covers, ignition system, intake manifold and carburetor or fuel injection, the exhaust manifolds, and the starter and alternator.

Weights listed for Lycoming and Continental aircraft engines are for items listed in the LONG BLOCK description, and should be compared thusly, not as ready to run, complete engines. There is only one accurate way to compare auto and aircraft engine weights; weigh each engine as it is ready to run.

Take a look at how far forward in the available engine compartment space this 200 hp Lycoming O-360 engine sits. A 350 hp Chevrolet V-8 engine that weighs 100 lbs. more could be mounted closer to the firewall and still stay in the C.G. range.

HEAVY AUTO ENGINES

At an Oshkosh Forum, one builder asked if there were heavy auto engines that should be avoided. Yes, for sure there are. A 460 cubic inch Lincoln V-8 engine weighs over 800 pounds and produces about 325 horsepower. It is far too heavy, even for a boat motor. Also, a big-block Chrysler or Plymouth V-8, 426 cubic inch engine is about as heavy as a Lincoln V-8. Avoid those engines. Read this chapter for more facts on engine weights.

Richard Richel, of Santa Barbara, California, built this 2 place Lancair 320, then installed a Lycoming with a big turbocharger and a 3 blade M.T. prop. The total engine and prop weight, firewall forward, was over 400 pounds. Richard won the 1997 Denver to Oshkosh Race, Corinthian Class in this airplane.

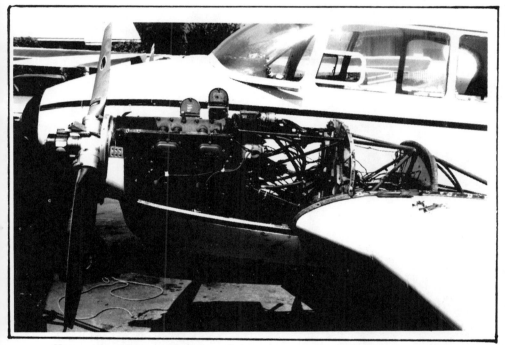

Many, many of us multi-engine pilots got our multi-engine licenses in Piper Apache airplanes like this one. Notice how far forward the 340 pound, 150 hp Lycoming and its 55+ pound featherable prop sit. You could easily mount a 325 hp Ford V-8 engine and 3-blade prop in all the room inside this engine cowl.

FAA PILOTS

If you and your co-pilot are the standard size, FAA defined five foot, nine inch tall, 174 pound people, and if you have not added a single extra bolt, rivet, or ounce of resin to your airplane, and if it weighs exactly what the designer calls for it to weigh, then you are the kind of person who should worry if your auto engine conversion weighs even one pound more than the plans for your airplane specifies.

.But, chances are, you are a few pounds heavier than the standard 175 pound FAA-specified pilot. Most builders and pilots are these days. Your author of this book has gained 45 lbs since his 35th. birthday. He (I) used to weigh 145 pounds and was six feet one inch tall. I constantly wished to be 175 pounds in weight with broad shoulders and big arms. I got my wish, but only weighed the perfect amount for a very few months. I can take the weight off now, but ice cream tastes soooo good! And so do peanuts, corn chips, and tacos. We all weigh too much these days...and so do our airplanes.

EXACT WEIGHTS

Trying to have your experimental airplane weigh exactly what the designer specified is a habit that began in the 1920's and 1930's when our engines produced 25 to 35 hp. The Wright Flyer engine produced only 12 hp. When you only have a few horsepower, you sure don't want too many extra pounds..

But when you have lots of horsepower to use, the extra horsepower will lift a few extra pounds of engine weight. That is true as long as your airframe is strong enough to support a few extra pounds of weight.

So because of these early day rules of aircraft building, many people still believe that your engine must weigh the exact same number of pounds that the certified aircraft engine weighs, or your airplane will not fly, or that it will not be safe. As this chapter progress, you will read and understand that these old-time beliefs are no longer true, because of modern engine technology.

NEW DESIGNS

Almost every time a new aircraft design is produced and flown, soon after the initial flights are completed, someone stuffs a bigger, more powerful engine in it.

The P-51 Mustang ended up with an engine that was twice as powerful and somewhat heavier than the first engine that powered it. My former employer, Mr. Ted Smith, built the first Aerostar twin-engine, 6 passenger plane and installed two 1O-320, 160 hp engines in it. Just months later, he built a second Aerostar and put two, six-cylinder, 290 hp 1O-540 engines in it, and it certainly did fly better. A few years later, he a converted an Aerostar to fly with two 400 hp IO-720 Lycoming

Sheril Dickey, of Phoenix, Arizona, resolved a C.G. problem in his E-Racer airplane by mounting his Chevy V-6, Buick V-8, Nissan V-6, and O-360 Lycoming engines on the C.G. of the Canard airplane, and then he ran a drive shaft and gear box to the fixed pitch prop.

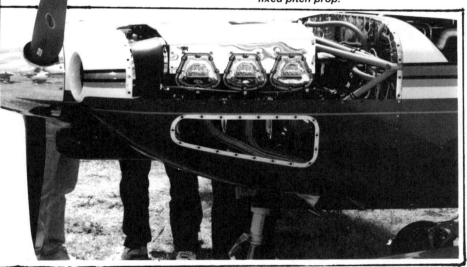

This SX-300, 2-place retractable is powered by a 600 pound Lycoming IO-540 engine with a 90 pound 3 blade prop. A big-block, 502 C.D. Chevy would fit here and weigh less, with a 3 blade composite prop.

engines for a total of 800hp, **and boy, did it ever fly better!!** From 160 hp, 300 lb. engines, to 400 hp., 700 lb engines, and the Aerostar just kept getting better and better. There were no negative results from putting bigger engines in the Aerostar (except cost and fuel consumption).

Most people remember the first 2-place Lancair that was flown, using a 200 cubic inch Continental, 100 hp engine taken from a Cessna 150 airplane. Recently, a friend of mine installed a 360 cubic inch, turbocharged Lycoming with a constant speed prop in his two place Lancair, and the 235 horses it produces really makes the airplane fly well. A picture of that airplane is shown on page 29 in this book.

What you have been reading in the past several paragraphs is that it is OK to put larger, heavier, more powerful engines in your airplanes, so long as you verify the C.G. (center of gravity), weight and balance, and the structural integrity of the air frame. But obviously, you would not want to put a 620 cubic inch V-8 in a plywood 2 place KR-2 airplane. There ARE certain limits.

GRUMMAN TIGERS

The 4 place Grumman Tiger airplane is certified with a 360 cubic inch, 180 hp Lycoming engine with a 32 pound fixed pitch, 2 blade aluminum prop. The complete firewall-forward installation weighs 375 to 395 pounds.

But a couple of men that I know have pulled out the 180 hp engine from their Grumman Tiger airplanes and have put in 6 cylinder, 260 hp, 540 cubic inch Lycoming engines and 65 pound constant speed props. The new engine installations weigh 690 pounds, firewall forward. Both men obtained STC's (supplimental type certificates) on the airplanes, and they are flying certified airplanes now!

Even Chevy cars and trucks save weight by going from 20 pound starters to 10 pound starters like this one on Jess Meyers Chevy V-6 in his RV-6A.

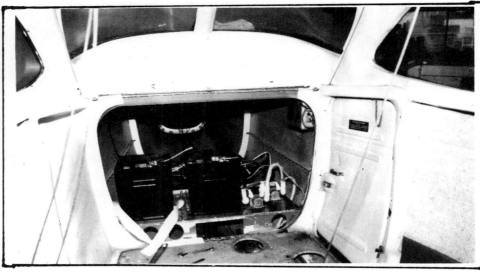

Elwyn Johnson, of Northwest Aero Products, Seattle, Washington, over-solved a C.G. problem when he installed a Chevy V-6 engine in his Cessna 172. He put 2 automotive batteries behind the baggage compartment, and then found that his now faster 172 was tail heavy!

The Jeff Ackland Legend airplane uses a Donovan Engineering, all-aluminum big-block like this one, more for the 620 cubic inch displacement than to conserve weight.

The point of telling you about this in this book, is to show you that a slightly heavier, but more powerful and much smoother running auto engine will work just fine in your airplane. The answer is that some auto engines are even lighter than the equivalent horsepower certified aircraft engine is, and in other instances, the auto engine conversion is slightly heavier than the same horsepower certified aircraft engine is.

For example, the Stratus Subaru EA-81 engine produces 115 hp, and it is 30 to 40 pounds LIGHTER than a 220 pound Continental 0-200 aircraft engine. And then, the 175 hp Belted Air Power V-6 Chevrolet engine weighs 45 pounds more than a 180 hp Lycoming. And a 300 hp Chevrolet V-8, also by Belted Air Power weighs over 100 pounds less than a 290 hp Lycoming IO-540.

FAA STANDARD PILOTS

At a recent Oshkosh Forum, I was speaking to a crowd of over 400 people who were interested in converting auto engines for experimental aircraft. I asked for a show of hands of those pilots and future pilots who weighed EXACTLY 170 pounds as specified in most light, general aviation aircraft information manuals. Two men raised their hands...out of an audience of over 400!

Next, I asked all those who weighed LESS than the FAA standard 170 pounds to raise their hands. Maybe 10 to 15 people raised their hands! Then I told the rest of the crowd of 400 plus that they could not fly or ride in airplanes! And of course I got a lot of applause and laughter because they all knew that there ways to get around the WEIGHT PROBLEM.

One way to get around the pilot weight problem if you weigh 220 pounds, is to put a 120 pound person in the co-pilot seat. If you are conscientious and do a proper weight and balance calculation before every flight, you will be able to shift the passengers and baggage around so that you stay within the weight and balance envelope. You can do the same weight shifting exercise when you are planning an auto engine installation in your experimental airplane. You can mount the battery in the rear of the airplane and you can move the engine closer to the center of gravity, regardless of whether the engine is in the front or the rear.

AUTO ENGINE WEIGHT

One of the many questions usually asked is about the apparent heavier weights of a 200-horsepower auto engine as compared to a 200-horsepower aircraft engine. Generally speaking, an auto engine with a reduction drive unit installed, complete with exhaust system, starter, generator, carburetor, and ignition, will weigh about 50 pounds more than its equivalent horsepower airplane engine.

This added weight is usually not significant because the auto engine is always mounted further aft in the engine compartment than the aircraft engine would be. This is because the usual prop speed reduction adapter places the mass or center of gravity of the auto engine closer to the firewall and, therefore, closer to the center of gravity of the airplane.

For instance, you can figure your weight and balance with a Lycoming 200-horsepower engine with its 325 lb.C.G. at +40 inches, and a V-6 auto engine with its 390 lb. C.G. at +30 inches, which places the actual engine weight 10 inches further aft than the aircraft engine. This might make the aircraft total empty weight 65 lbs. heavier, but cause the aircraft C.G. to stay the same or even move aft with an auto engine in place of an aircraft engine.

As an extreme, but obvious possibility, you could spend an extra $7,000 to buy aluminum cylinder heads and an aluminum short block, rather than a standard 4.3 liter Chevrolet cast iron V-6 engine. When you assemble this all-aluminum Chevrolet V-6, it would weigh almost 100 pounds less than a 200-horsepower Lycoming. Then you add a 50-pound gearbox to the Chevrolet V-6, and it still weighs 50 pounds less than a 200-horsepower aircraft engine!

So, you must ADD lead to the nose of the airplane to bring the center of gravity back into range! So, you just wasted $7,000!

ALUMINUM ENGINES VS. CAST IRON ENGINES

Are they really worth the cost? Or, would you spend $2,650 to save 48 pounds? That is how much more a cast iron Chevrolet V-8 engine block weighs than an aluminum Chevrolet V-8 engine block does.

A basic 350-cubic-inch Chevrolet V-8 engine block will weigh about 100 pounds. That figure includes the main bearing caps, main cap studs, and core plugs. The same configuration in a cast iron Chevrolet V-8 engine block will weigh about 175 pounds, but with selective machining to remove unneeded lugs, bosses and dead weight, the cast iron V-8 block can be pared down to 148 pounds.

Cast aluminum cylinder heads weigh 18 pounds less each than cast iron Chevrolet V-8 cylinder heads. You can save 36 pounds by spending $600 extra for aluminum cylinder heads for the Chevrolet V-8. This is a more reasonable way to save weight if you simply MUST save weight.

To simplify the equation of dollars to pounds, it costs $55 to save a pound in the engine block but only $17 to save a pound in the cylinder heads. And you can actually save too much weight when converting auto engine power in your airplane. For instance, if you are converting a Lancair 4 airplane from the 350-horsepower Continental six cylinder, twin-turbocharged engine to a 350-cubic-inch, 350 horsepower, twin-turbocharged Chevrolet V-8, you will actually have to ADD weight to the auto engine to make it as heavy as the Continental!

To give an example, Florida race car engine builder Smokey Yunick took a small block Chevrolet to the 1973 Indianapolis 500 race, and that particular cast iron small-block engine weighed only 505 pounds, complete with two turbochargers. I weighed a Continental 6-cylinder C-GTSIO-520-K, 340-horsepower aircraft engine in 1979, and it weighed 670 pounds! You would have to ADD 165 pounds of lead to the nose of the Lancair 4 airplane when

installing the Chevrolet V-8 in place of the Continental six-cylinder.

Another BIG difference is that the Chevrolet V-8 was capable of 1,000 horsepower output, whereas the Continental would blow up if you tried to extract an extra 100 horsepower from it. Auto engines are simply better and more dependable than aircraft engines. No doubt about it.

WEIGHTS OF SMALL AUTO ENGINES

If your airplane needs far less horsepower than the 350-cubic-inch Chevrolet engine produces, you might want to consider a V-6 engine. But when you buy aluminum cylinder blocks and heads for a V-6 engine, the pounds of weight you save is 20% less in most cases, and the cost is 50% more. To lighten a 4.3 liter Chevrolet V-6 engine by 50 pounds, you must spend at least $3,500 for the cylinder block, and $2,000 extra for the cylinder heads. This extra cost figure of $5,500 will save you 50 pounds, at a cost of $110 per pound saved. So, you must ask yourself if it is worth the cost to save a few pounds.

WEIGHT X ARM = MOMENT

Or, you could call this the weight-and-balance theory. It works with anything that needs to have a balance point calculated. We need to know where the engine weight (center of gravity and concentration of weight) needs to be placed. We can calculate this distance and C.G.

So, for a simple explanation, let's imagine two kids on a teeter-totter, or-as we call it in Texas-a see-saw. Children playing on a see-saw illustrate the basic principles involved in weight and balance theory. For airplane purposes, where the wing is the fulcrum point, and is not moveable, let's assume the see-saw is fixed in the middle (to its support point) and that the children may move along the arm, or length, of the see-saw to balance it.

If the see-saw is 200 inches long and is supported in the middle, the seats can each be 100 inches from the middle if both children weigh the same. But if one child

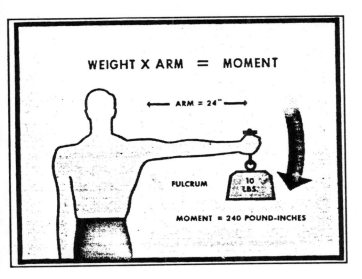

WEIGHT X ARM = MOMENT

ARM = 24"

FULCRUM

10 LBS.

MOMENT = 240 POUND-INCHES

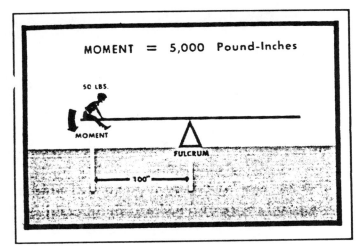

MOMENT = 5,000 Pound-Inches

50 LBS.

MOMENT

FULCRUM

100"

One boy sitting on a see-saw represents a weight-and-balance problem. You can relate airplane balance problems to this illustration.

leaves, and a heavier child comes to sit on the see-saw, he must sit closer to the support point of the see-saw in order for the see-saw to balance.

5,000 POUND-INCHES

The kid on the left side of the see-saw weighs 50 pounds, and is sitting exactly 100 inches from the center balance point of the see-saw. We can calculate his moment arm by multiplying his weight times the arm (or distance from the balance point), 100 inches, and the answer is 5,000 pound-inches. Keep that number in mind.

ADD 50 POUNDS

When a heavier kid gets on the other end of the see-saw, he will weigh it down because he weighs 100 pounds (multiplied times 100 inches) and his moment arm is 10,000 pounds. So, to balance the see-saw, we can have him move toward the center until his 100 pound weight times an unknown distance from the center of the see-saw equals 5,000 pound-inches, which is the same moment arm as the 50-pound child. Simple calculation shows that the 100-pound child needs to move to a 50-inch point on the see-saw in order to balance both children with equal moment arms of 5,000 pound-inches. The way we do this is to divide the desired 5,000 pound-inches by the child's weight (5,000 divided by 100 = 50 inches).

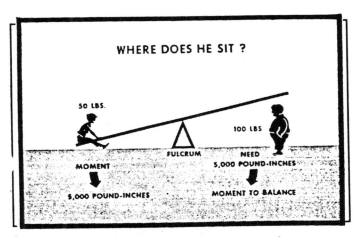

WHERE DOES HE SIT ?

50 LBS.

100 LBS

FULCRUM

NEED
5,000 POUND-INCHES

MOMENT

5,000 POUND-INCHES

MOMENT TO BALANCE

Here a 100-pound boy is ready to sit on the see-saw. Where does he sit in order to balance the see-saw? Heavier airplane engines represent a similar problem.

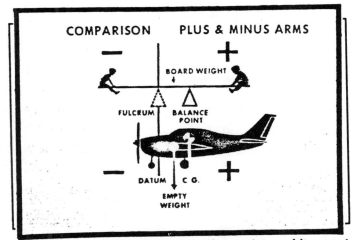

COMPARISON   PLUS & MINUS ARMS

BOARD WEIGHT

FULCRUM   BALANCE POINT

DATUM   C.G.

EMPTY WEIGHT

In an airplane, the wing is the fulcrum point, and it cannot be moved, so you must calculate where to place a heavier engine or a lighter engine.

ENGINE WEIGHTS

Now that you understand how shifting weight nearer or further along the moment arm makes the see-saw balance, you can easily calculate the same thing for your airplane.

If your 200-horsepower Lycoming engine weighs 340 pounds, and its center of gravity is 52 inches from the C.G. reference point, it has a moment arm of 17,680 pound-inches, similar to the child on the see-saw. When you add an auto engine that weighs 410 pounds, you simply divide the 17,680 pound-inches of your airplane engine by the auto engine weight, and this gives you the desired location, fore and aft, of the auto engine installation. This equates to 43.12 inches ahead of the C.G. of the airplane, or 8.88 inches aft of the Lycoming engine location.

Because the auto engine gearbox causes the mass of the auto engine to be closer to the airplane firewall, it is seldom necessary to add ballast to the tail to offset the extra 70 or so pounds of the auto engine. "But," you say, "what about the extra 70 pounds that must be added to the basic weight of the aircraft?" There are at least two answers to this question.

CARRY LESS FUEL

A 200-horsepower, 340-pound Lycoming aircraft engine will burn about 10 gallons of fuel per hour at cruise power. That equals 60 pounds of fuel burned per hour. If your airplane fuel tanks hold 60 gallons of fuel, you can fly six hours before your tanks run dry. That equals about 960 miles if your airplane cruises at 160 MPH.

But a 410-pound V-6 auto engine, complete with a prop speed reduction drive unit, will produce about 250 horsepower, and burn about 7 gallons of fuel per hour, while cruising at least 20 miles per hour faster. So your 70-pounds heavier auto engine will take you 960 miles in 5.3 hours at 180 MPH, and it will only use 37 gallons of fuel during the flight. So, you can put 23 gallons less fuel in the tanks, and

start your trip with 138 pounds less fuel. This means that you are actually 68 pounds lighter with the auto engine than you were with the obsolete aircraft engine.

TELL THE PILOT TO GO ON A DIET!

Seriously, it might be easier for 220-pound pilot to lose 40 pounds, down to 180 pounds, than it would be to cut 40 pounds off a car engine installation. The cost per pound certainly would be less.

CALCULATING FOR A HEAVIER AUTO ENGINE; AN EXAMPLE:

You will want to relocate your heavier auto engine center of gravity closer to the firewall to keep the airplane C.G. within the designer limits. We will use an example of installing a 276 pound engine where a 158 pound engine was once installed. You can adjust these figures to correspond with the weights of your aircraft and engine.

#1. Compute the difference in engine weights. New engine is 276 lbs., old engine is 158 lbs. Difference is 118 lbs.

#2 Solve for adverse moment: difference = 118 lbs. x engine C.G. to datum, - 29" = adverse moment - 3,422 " lbs.

#3. Compute distance to move engine toward firewall - 3,422 lbs. divided by 276 lbs. = -12.4" move aft.

Practical example:
Old aircraft engine wt. = 345 lbs.@ 29.0"
New auto engine wt. = 405 lbs.
Difference is 60 lbs. x -29" = -1,740 lbs.
1740 " lbs. divided by - 405 lbs. = 4.29" aft.

In other words, the cast iron auto engine would only have to be moved aft 4.29". In reality, the auto engine C.G. would be further aft by design of the engine anyway. The most important part of this calculation is to find the actual C.G. of your auto engine AND prop speed reduction unit.

Pictured here is a cut-away of a 1998 Chevrolet Corvette LS1 all aluminum engine that is only 10 pounds lighter than a cast iron Corvette engine of equal power.

TO MODIFY OR NOT

◆◆◆◆◆◆◆◆◆◆◆◆◆◆◆◆◆◆◆◆◆◆◆◆◆◆◆◆◆◆◆◆◆◆◆◆

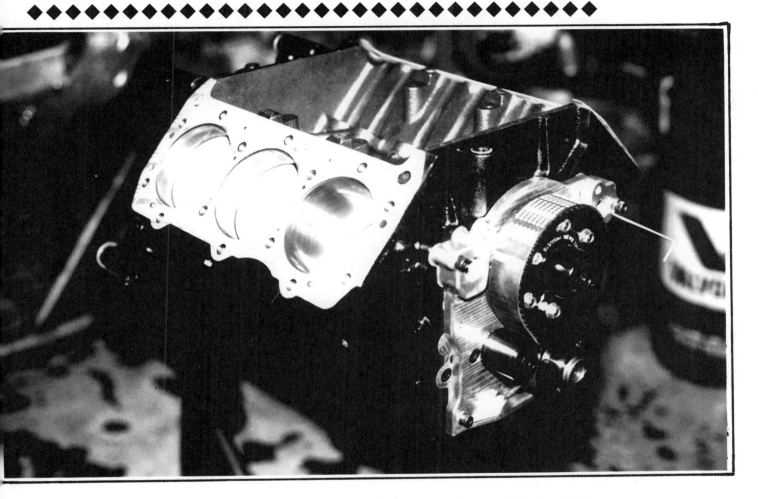

This 3.8 Liter, stage II Buick V-6 engine short block is being built up to produce over 1,000 hp with twin turbochargers and using a giant size intercooler. The cog belt timing "gear" is used to produce more precise valve-to-piston timing. Ken Duttweiler High Performance Engines, of Saticoy, California assembled this bullet-proof engine.

TO MODIFY OR NOT?

YES!

Even if you merely strip all the excess accessories off your choice of auto engines to convert for your airplane project, stripping the engine down is a form of modifying it.

You will also need to remove the cast iron exhaust manifold, the power steering pump and its related bracketry, the air conditioner compressor and brackets (unless you want air conditioning in your cross country airplane), and on many engines removed from low mileage salvaged cars, you will need to remove the decorative trim pieces that the automotive manufacturers put on the engines to make them look attractive. But you should not make major modification to the engine, in most cases.

The bottom end of the 3.8 Buick V-6 engine shown on page 35, is fully balanced and features a special racing version of the even-fire crankshaft. Notice also that the two center main bearing caps have provisions for 4 bolts rather than the 2 stock bolts.
Photo: Ken Duttweiller Automotive

ENGINE CHOICES

The amount of modification you will want to do to the engine depends almost entirely on the engine you have chosen to convert for aircraft use. The opposite examples of engine choices might be a factory rated 200 hp V-6 engine for your homebuilt airplane that flies well on 180 to 200 hp, and the opposite example might be a 502 cubic inch, 502 hp big block V-8 for your replica P-51 Mustang or your 8 place custom built bush plane.

And there are many choices in between and below those horsepower figures and even above those horsepower figures. Many people have found the need for 50 hp car engines for their ultralight aircraft designs, and still other auto engine converters are finding the need for V-10 and V-12 engines of 1,000 horsepower and above.

Auto engine conversions have successfully flown with cash outlays of as little as $300, and others with expenditures of over $100,000. As the interest in auto engine conversions grows, more and more builders want the simplicity, the dependability, the smoother operation, and the much lower cost of equivalent horsepower in an auto engine versus the much

higher cost of an antiquated design certified aircraft engine.

In the following paragraphs, this book will explain how much to modify the engine you have chosen. This chapter will explain how much to modify the engine you have chosen. This chapter will examine various engine choices according to their cost, age, horsepower output and complexity of design.

INCREASES OVER FACTORY HORSEPOWER

Because all modern day automobiles (1989 & later) are powered by engines that have been engineered to reduce exhaust and crankcase emissions, the actual horsepower outputs are somewhat less than the horsepower that is possible from a certified aircraft engine of similar advertised horsepower. If the world governments decide to require smog and emission equipment on general aviation piston engines, you can expect that your 200 hp certified aircraft engine will only produce 150 to 160 hp when it is "emission legal".

What this means to you is that a 2.7 Liter, 165 cubic inch auto engine can produce 165 to 175 horsepower when you eliminate the

exhaust catalytic converter, EGR exhaust gas recirculator, the restrictive intake and exhaust systems, and the systems that reduce emissions during low power, stop and go city driving in all climates, from below zero winters to 125 degrees F. and higher in desert summertime driving.

When these modifications are made, additional horsepower is possible and available, a typical V-6, V-8, or V-10 auto engine can be "hopped up" to produce at least one hp per cubic inch of displacement.

OLDER ENGINES

A lot of interest still exists in converting aluminum V-8 engines made in 1961 to 1964, namely the 215 cubic inch Buick and Oldsmobile V-8 engines. You should be aware of the fact that these engines are actually heavier than newer 231 cubic inch V-6 Buick engines that are cast iron. The reason for the heavier weight in the older aluminum engines is the weight of the two extra cylinders and the necessity to make the aluminum castings heavier to equal the strength of cast iron used in newer engines.

Another thing to be aware of in converting older, high mileage engines, is the characteristic of metal to fatigue when stress reversals (power strokes) are added up into the tens of millions of strokes. The FAA requires that a certified aircraft engine be able to withstand ten million stress reversals in order to pass the certification tests. In a highway driven automobile, that equals less than 42 hours on the highway, and less than 2,300 miles of driving! An older engine with 100,000 miles or more on it will be very subject to fractures from metal fatigue. One well known auto engine converter experienced a broken connecting rod in his 30 year old Buick V-8 engine, even though he polished the rods to remove surface cracks. Therefore, if you decide to convert one of the older engines, you should consider buying all new internal parts such as the crankshaft, rods, pistons, piston pins, rod bolts, main bearing bolts, and especially all the parts in the valve train.

Considering these costs, you would want to investigate the costs of converting a late model V-6 Buick (Pontiac, Oldsmobile, etc.) that weighs less and that is not weakened by high mileage.

NEW CRATE ENGINES

As noted elsewhere in this book, the most horsepower per dollar that you can buy is to go to your General Motors dealership and buy a brand-new (NOT rebuilt) Corvette V-8 engine for $2,695 to $2,995, completely assembled and ready to bolt right into your car, boat, or airplane. The modifications you would want to make to this engine would be to change the water pump to an aluminum pump to save weight, change the crankshaft pulley to a smaller one to slow down the water pump, the alternator and other belt-driven equipment to a decent rpm when the engine is cruising along at 4,000 rpm in the sky. You should also consider changing the alternator and water pump pulleys to

larger diameter units to further reduce the rpm's of the water pumps and alternator at 4,000 rpm of the engine at cruise.

You will also want to consider adapting an oil filter fitting that will provide for an oil cooler and a remote oil filter.

Look in the PA&W auto parts catalog for oil filter and oil cooler adapters. They are available in sizes to fit almost any auto engines. The cost of these adapters is usually about $50.

This is a 425 hp stage II Buick V-6 engine that is used in the Indy Lights race cars. Several MILLION miles of racing without engine failure proves that this 4.1 Liter engine is very dependable at high rpm and high power output.

Ken Duttweiler drives this much modified Buick V-6 to 180 mph at only 7+ seconds in the 1/4 mile. The big radiator in the front is an air-to-air intercooler to cool the turbocharged inlet air. Ken Duttweiler Automotive has dynoed this engine at 1,300 hp @ 9,000 rpm.

MODIFY IT YOURSELF?

If you KNOW for sure that you are capable of overhauling or re-building the engine in your family car, then you should not have any problems in rebuilding an auto engine for aircraft use. But if you doubt your engine building skills, you probably should leave this task to someone you trust to do this critical job properly.

Of course, there is always the first time for every competent auto engine overhaul mechanic. If you are willing to ground test your first auto engine overhaul for enough time to prove its dependability, then you will likely be successful. The FAA rules and regulations say that 150 hours is sufficient time to certify a new design aircraft engine. You could cut that time by 90% and ground test, including high speed taxi tests, and fly the engine with as little as 15 to 20 hours of ground test time.

DESIRED MODIFICATIONS

If you decide on using a high mileage used engine from a wrecked or salvaged car, you will want to disassemble it completely and inspect it for unusual wear and especially for signs of overheating. If you can detect a strong sulfur smell inside the engine and if you find scored cylinder walls, a scorched crank shaft or heavy sludge in the inside of the engine, don't spend any more time working on it. Go trade the pieces back to the salvage yard and look for a better engine.

Once you have found a suitable engine, it should be modified as follows:
1. Bore the cylinders to .030" oversize and install new forged (NOT cast) pistons.
2. Regrind the crankshaft to .010" undersize and install new rod and main bearings.
3. Line bore the block for new cam bearings and new main bearings.
4. Do a three angle valve job and match the intake ports in the head to the intake manifold by grinding away excess metal.

5. Have the engine balanced, including the flywheel and any inertia wheel.
6. All work should be done by using the engine manufacturers' service manuals.

Obviously, the modifications suggested in the previous paragraph are only possible if you have access to a modern and complete automotive machine shop.

This 620 cubic inch Donovan block engine in Jeff Ackland's Legend airplane is modified to race car specifications, including a dry sump oil system for inverted flight. Note the large, remote oil sump at the firewall.

Jeff Ackland, under the engine at the left, has made all the systems in his engine race car quality or better, including the use of Grade 8 nuts and bolts throughout the engine compartment.

IF IT AIN'T BROKE, DON'T FIX IT!

This popular statement sure does apply to auto engine built since 1980. And it also applies to most low mileage, less than 50,000 mile auto engines manufactured since 1975. Stress reversals, which are firing strokes, do in fact affect the ultimate life of a piston engine. On the average an auto engine sees one million stress reversals, piston strokes, every 229 miles, and Ten Million stress reversals every 2,292 miles. In 50,000 miles, the auto engine sees 218 million stress reversals. Obviously, a high mileage engine would begin to show stress cracks in pistons and connecting rods. Therefore, avoid converting high mileage engines.

For experimental aircraft builders who want the much lower cost and obvious benefits of the dependability of auto engine power, the best engine choice would be a new engine in a crate or a very low mileage engine out of a salvage yard car. If this is your situation, look for an engine with less than 25,000 miles on it. Ask to see the car it came out of so you can personally verify the condition of the car and the odometer reading. And then resist the urge to tear it apart.

It is advisable to change to lightweight header type exhausts, a lightweight aluminum intake manifold if your engine does not already have one, and change the camshaft if your engine is more than 5 years old. But do not take the pistons out of a low mileage engine, just because.

THE MOST RELIABLE ENGINE

In this book you will see many examples of auto engines that are flying dependably. The most inexpensive and most dependable engine pictured in this book is the 4.3 liter Chevrolet V-6 engine in the Belted Air Power RV-6A airplane. If you are in the market for 180 to 200 horsepower, that engine is the way to go.

Tracy Crook of Florida installed a Mazda 13 B rotary engine in his RV-4 airplane, and incorporated three side draft motorcycle type carburetors to feed the two rotors.
Photo: Tracy Crook

The Belted Air Power engine was purchased new, in a crate, and SLIGHTLY modified. They changed the camshaft and roller lifters to a street rod cam, they changed the intake manifold to a milled down aluminum unit, and the exhaust system to a tubular steel rod type exhaust, and the ignition system to a dual point, dual coil system.

Other dependable engines might be the 350 cubic inch Chevy V-8 crate engine shown on page 26, or a Buick 3800 V-6 engine, fuel injected or even supercharged. Ford and Chrysler also have brand new crate engines that would be suitable and dependable.

ALL ALUMINUM KIT ENGINES

Ford, Chevrolet, Buick and Chrysler all have racing versions of their V-6 and V-8 engines. You can save 60 pounds by changing to an aluminum Chevy V-8. The cast iron Chevy V-8 engine block weighs 160 pounds and the aluminum Chevy V-8 weighs 100 pounds. If you elect to go to this type engine, you will need a DONOR ENGINE to rob all those little unavailable parts from. The same donor engine suggestion applies to PA & W (see page 152) kit engines.

PA&W ALUMINUM V-8'S

If your airplane can handle 350 horsepower to 425 horsepower, and you want to spend $10,000.00 for an all aluminum V-8 engine, you can purchase a complete kit from Performance Automotive Warehouse (PA&W) in Chatsworth, California. In the current PA&W catalog, there is an all aluminum 350 cubic inch small block V-8 engine kit listed for $9,000.00. They also list a big block Chevrolet V-8, all aluminum engine kit (427 cubic inch) for $10,000.00. But remember that you need a donor engine to get all those unfurnished parts that do not come in the kit.

OTHER ALUMINUM V-8'S

As pictured in this book, there are several other sources of aluminum small block Chevrolet and big block Chevrolet engines. Check on Rodeck, GM, and other brands for small blocks. DONOVAN is a good, cooperative source for aluminum big blocks.

For the lower budget aircraft conversions, you should strongly consider one of the GM crate engines that are listed in this book. These engines all sell new for $1,000.00 to $4,500. And these crate engines do not require much in the way of modifications.

This very compact and light weight Mercedes V-8 engine produces over 800 horsepower, but must rev to 14,000 rpm to do it. You are looking at an Indy race car engine out of race car driver, Al Unser, Jr.'s Indy Race car. That is too much modification for your experimental airplane.

Another 800 plus horsepower V-8 engine is this 2.2 liter (or less) Honda V-8 engine that came out of race car driver Jimmy Vasser's Indy Race car. Even if you could design and build a 7 to 1 reduction ratio gear box and adapt it to your airplane, remember that the TBO on this engine is about 5 to 10 hours at 14,000 rpm.

COMPARISON CHART -	LYCOMING	BUICK V-6	RENAULT V-6	MAZDA
HORSEPOWER	150 @ 2750 rpm	150 @4800 rpm	145 @ 5100rpm	145 @ 7000 rpm
TORQUE, ft lbs	Not listed	210 @ 2000 rpm	136 @ 2750	130 4500 rpm
CU. IN. DISP.	320 cu. in.	231 cu. in.	174 cu. in.	80 cu. in.
CU. CC DISP.	5,264 cc	3800 cc	2800 cc	1300 cc
WEIGHT	348 Lbs.	350 Lbs.	337 Lbs.	265 Lbs.
LENGTH	32 1/4 in.	24 1/4 in.	21 1/4 in.	19.0 in.
WIDTH	32 1/4 in.	19 3/4 in.	28 3/4 in.	19 1/4 in.
HEIGHT	27 1/2 in.	26 3/4 in.	27 1/2 in.	22.0 in.
ENG. TYPE	Flat 4 Cyl Opp.	V-6	V-6	Rotor
BLOCK MAT'L	Aluminum	Cast iron	Cast alum.	Aluminum
HEAD MAT'L	Aluminum	Cast iron	Cast alum.	Cast iron
FUEL SYSTEM	Carburator	Carburator	Bosch F.I.	Carburator
CAM SHAFT	Single	Single	Dual	None
CAM DRIVE	Gears	Chain	Belt	None
IGNITION	2 Magneto	1 H.E.I. Dist.	1 Electronic	1 Dual Ignition

Engine Mount Designs

◆◆◆◆◆◆◆◆◆◆◆◆◆◆◆◆◆◆◆◆◆◆◆◆◆◆◆◆◆◆◆◆◆◆

The early development stage of Belted Air Power's V-6 Chevy installation into an RV-6A airplane included this elaborate work stand to position the engine exactly where it would fit the airplane firewall the best. In this picture, you can see that the engine mount has been designed, developed and welded, and that the engine mount incorporates the nose landing gear fitting.　　Photo: Dr. Bill Harrold

ENGINE MOUNT DESIGNS

The first step toward making a successful auto engine conversion for your airplane (once you have selected an engine and PSRU), is to design an engine mount and bolt the engine to the airplane firewall. All of the engine systems such as intake, exhaust, ignition, cooling and cabin heat, will have to be routed around the engine and its structural mount.

TRIANGULATE

You can see several good triangles in this Jess Meyers Belted Air Power engine mount design on this page. This mount is very strong, and it works very well. The more equal the sides of a triangle, the stronger the structure will be.

A factory built "bed mount" for a 6 cylinder, geared Lycoming engine, is a very strong and adequate engine mount design. With a few changes of the dimensions this would be a good design for a V-6 or a V-8 auto engine in this same airplane.

DESIGNING THE MOUNT

There is a list of proven engine mount tubing sizes in this chapter. Once you have decided on the auto engine that you will be using, you can select the tubing sizes from the accompanying list. Most engines mounts require 80 to 100 feet of tubing. If you do a good job of designing the mount, and if you know that you are not going to make any mistakes in building the mount, you may be able to get by with 60 to 70 feet of tubing.

There are three ways to design an engine mount, and all three methods require that you build the mount in a 3 dimensional, full size prototype before you build the flightworthy mount.

METHOD ONE

If you can get a 3-view drawing, to scale, of your engine, the best design method is to do a rough, lay-out sketch of the mount as is shown in this chapter. Even the "certified " aircraft factories start with such a sketch. If you are a very good draftsman or a CAD operator, you can even draw a detailed engine mount design.

Try to keep all the angles in the mount as equal as possible. The strongest framework is one that has equal side length triangles in the design. Steep angles in the design have less rigidity and less strength. After the engine mount is designed on paper, the next step is to build a prototype mount quickly, from disposable materials. You can glue a prototype mount together, made from PVC plastic water pipe.. Another method of making a prototype mount would be to braze up a mount, using 1/2", 3/4", or 1" emt galvanized electrical conduit tubing.

A third method would be to actually weld up the mount from 4130 steel tubing, fit checking the engine as you construct the mount.

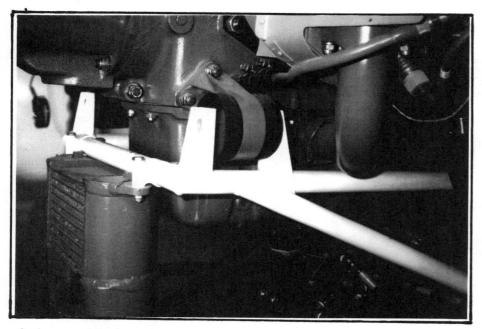

A close-up of the Lycoming "bed mount" that uses a $100 rubber isolator. Auto suspension bushings do the same work for $14 each. Try an automotive part number MOOG #K-5274, and build your engine mount like this one.

PROP OFFSET

In older, slower airplane designs, most notably the 85 mph Ercoupe, the engine thrust line is noticeably down and to the right, to offset prop "P" factor. But this offset slows the airplane in cruise flight. Most newer airplane designs place the engine thrust line directly ahead, so the airplane will fly faster.

One of the hardest parts about installing an auto engine in an experimental airplane is designing and building a safe, suitable tubular steel engine mount to support the engine. The photos and sketches in this chapter will show you how to do it.

Once you build and install a safe engine mount, you can hang the auto engine in the mount, on the airplane, and then you can really make good progress with your conversion.

FINDING YOUR ENGINE'S C.G.

This little, but very important, job is very easy to do if you have one of the 2-chain yokes that are used to install engines into cars. Simply hook the chains from the yoke on the face of the PSRU prop flange, and the other chain somewhere near the extreme end of the engine where the pulleys fit up against the firewall of the engine.

Then lift the entire engine assembly off the floor, bench, or wherever it is resting. If you have attached your chain or cable lift in the middle of the yoke, one end of the engine will tilt up noticeably.

Simply place a level on a flat part of one of the valve covers if the engine is a Vee Type, and crank the yoke to support the down - or heavy - end of the engine until the engine hangs level, front to rear.

Then, draw an imaginary line straight down from the lifting cable, and that is the center of gravity of your engine. Mark it permanently, with paint. Record the C.G. location with a photograph for future information.

Now you are ready to figure your engine C.G. and position ahead (or behind) the firewall, and determine how the engine mount should be built.

ENGINE MOUNT TUBING SIZE

- For 350 hp. V-8 engines weighing 600 lbs. or less, use 1 inch O.D. by .063" wall thickness 4130 steel tubing.

- For 200 hp. V-6 engines weighing 425 lbs. or less use 7/8 inch O.D. by .049" wall thickness 4130 steel tubing.

- For 150 hp. V-6 & 4 cylinder engines weighing 300 lbs or less, use 3/4 inch O.D. by .049" wall thickness 4130 tubing.

- For 100 hp. 4 cylinder engines, weighing 200 lbs or less, use same tubing as for 150 hp engines.

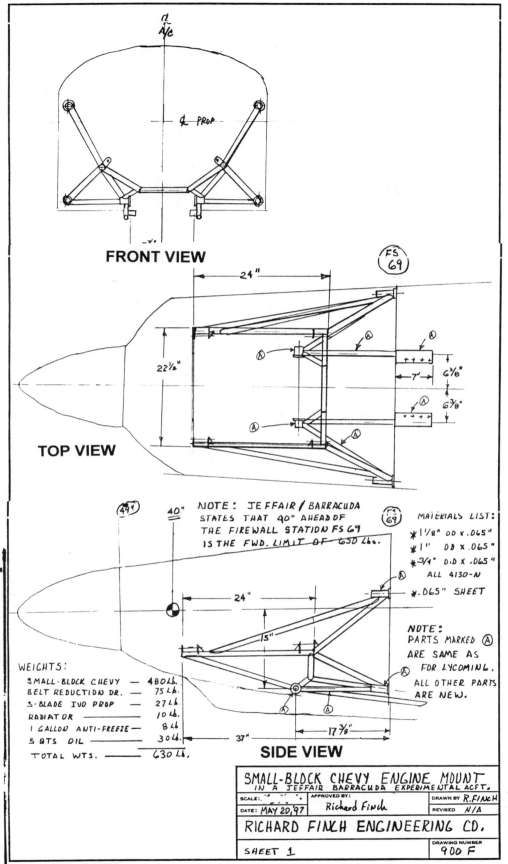

FRONT VIEW

TOP VIEW

NOTE: JEFFAIR / BARRACUDA STATES THAT 40" AHEAD OF THE FIREWALL STATION FS 69 IS THE FWD. LIMIT OF 650 Lbs.

MATERIALS LIST:
* 1 1/8" DD x .065"
* 1" DD x .065"
* 3/4" D.D x .065"
  ALL 4130-N
* .065" SHEET

NOTE: PARTS MARKED Ⓐ ARE SAME AS FOR LYCOMING. ALL OTHER PARTS ARE NEW.

WEIGHTS:
SMALL-BLOCK CHEVY — 480 Lb.
BELT REDUCTION DR. — 75 Lb.
3-BLADE IVO PROP — 27 Lb.
RADIATOR — 10 Lb.
1 GALLON ANTI-FREEZE — 8 Lb.
5 QTS OIL — 30 Lb.
TOTAL WTS. — 630 Lb.

SIDE VIEW

SMALL-BLOCK CHEVY ENGINE MOUNT
IN A JEFFAIR BARRACUDA EXPERIMENTAL ACFT.

SCALE:	APPROVED BY: Richard Finch	DRAWN BY R.FINCH
DATE: MAY 20, '97		REVISED N/A

RICHARD FINCH ENGINEERING CD.

SHEET 1	DRAWING NUMBER 900 F

8½ X 11   PRINTED ON NO. 1000H CLEARPRINT ®

This sketch adapts a Lycoming IO-540 engine mount from a Jeffair Barracuda airplane to accept a 350 hp Chevrolet V-8, Finchbird belt drive and 3 blade electric IVO-prop. Total weight will be 600 pounds, nearly 100 pounds LESS than the 290 hp Lycoming and 3 blade constant speed prop. And nearly $40,000 less?

METHOD TWO

If drawing detailed plans is not your cup of tea, you can support the engine exactly where it should fit on the firewall, and start measuring and tack welding tubing in place. This method works fairly well if you can MIG (wire feed) tack the mount at the engine and the firewall, or if you can TIG (heliarc) tack the tubing. But if you are having to tack weld with oxyacetylene gas welding, you will likely have trouble with the gas flame scorching or burning the paint off your engine and airplane.

I built my first engine mount, to install a Renker Gear Drive, Turbocharged Corvair engine in a 1960 Cessna 150 airplane, by hanging it in the proper location in the engine compartment and gas tack welding the mount together. After it was tack welded, the mount was taken to a friend who TIG welded it in its fixed configuration.

This method works if you are really careful to not catch your airplane on fire with the gas torch flame, but it would work a lot better if you have a TIG welding outfit in your shop.

METHOD THREE

A significant number of designers and builders prefer to hang the engine and PSRU vertically over a plywood outline of the firewall of their airplane. This method works well with certain engines such as the Subaru EA-81 and Legacy, and the Mazda Rotary engine, because the center of gravity of those engines are nearly through the center of the crankshafts. What this means is that the engine will hang close to vertical and it only requires slight brazing and blocking up to properly position it in relation to the line of thrust on the firewall pattern.

Obviously, if you have selected a big block Chevrolet V-8 engine with a 10 inch offset gearbox or chain drive PSRU, it would be very hard to hang the engine from its prop flange and make it hang straight. The best way to hang a V-6 or V-8, V-10 or V-12 engine, is to hang it by each end and find its C.G. before you build the mount for it (see

Jay Blair of Cannon Falls, Minnesota, is using 3/4 inch wood dowels to mock up an engine mount for a 4.3 Liter Chevy V-6 in his Glassair RG airplane. Photo: Jay Blair

The author mocked up an engine mount for his 3.8 Buick V-6 engine and gearbox, by brazing a mock-up mount out of inexpensive 3/4" EMT electrical conduit. No down thrust or right offset was used in this mount design.

Page 43) This will help you calculate your aircraft C.G. and tell you how far ahead of the firewall the engine should be positioned. This is a very important step in engine mount design.

Then hang or mount the engine in front of the firewall or a firewall mock up and weld up a mock up mount in place.

One very convenient thing about hanging a center-crankshaft engine such as a Mazda rotary, or a Subaru, over a firewall mockup is that it will be very easy to see the bottom of the engine where the majority of the engine mount goes. It is easier to build a mount in place this way than any other way.

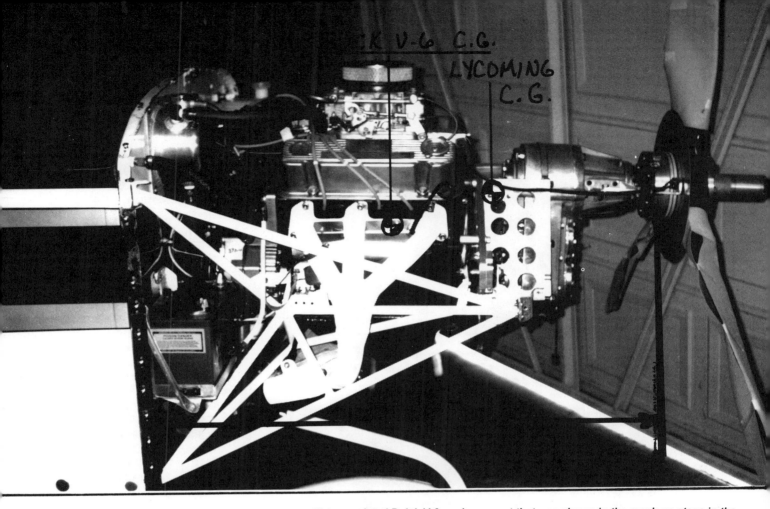

The completed Buick V-6 engine mount that was shown in the mock-up stage in the previous picture, makes use of the automotive suspension bushing, MOOG #K-5274. In this picture you can see the 3-blade electric pitch change IVO-Prop. Also, notice the apparent center of gravity of the engine is about 2 inches ahead of the center exhaust port on the engine.

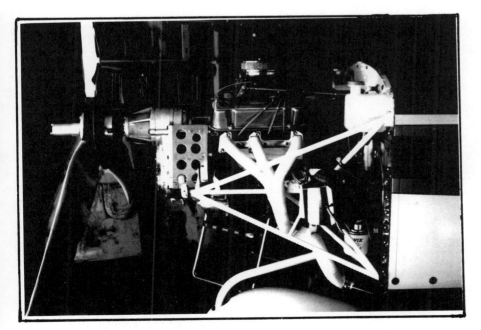

This left side picture of the Buick V-6 engine mount shows how the exhaust header is routed between the engine mount tubing. You can also see the end of the muffler and the chrome exhaust tail pipe tip.

WEIGHT SHIFTING

If you want to install a cast iron Chevy or Buick V-6 engine in your two-place experimental airplane, it can surely be done without any negative side effects. You simply move the engine nearer the center of gravity of the airplane, by as much as 10 or 12 inches. If you are installing the V-6 engine in a pusher airplane, the engine can be moved forward toward the center of gravity of the airplane. If you are installing the engine in a tractor-type airplane like a Lancair 360 or a Glassair TD, you move the engine aft a few inches. This C.G. placement is determined by the

engine mount you build. See Chapter 5 for more information.

The engine mount shown in this picture is a Fred Geschwender unit, supporting a Ford V-8 with a Morse HY-VO chain drive in a 7/8 scale Venture Mustang, circa 1975. This is an excellent design and it worked very well.

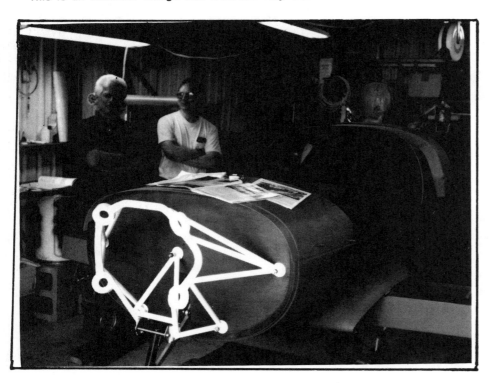

John White, left, and John Blackwell, right, consider modifications to the LYcoming mount on the front of their matching Lancair 235's, to accept 60 degree V-6 Chevrolet engines.

In order to change from Lycoming or Continental engine power in many experimental airplanes such as this Lancair 235, designed for a 235 cubic inch, 115 horsepower Lycoming engine, you must design and build a new engine mount that makes provisions for the firewall mounting points, and for the retractable nose landing gear mechanism. But don't let this design problem stop your efforts. Remember that most equal horsepower auto engines are actually more compact than the certified aircraft engine that they replace, and the engine mount can always be re-designed to accept the auto engine.

PRACTICAL APPLICATION

Next, we will take a specific engine installation in a specific airframe and do the engine location calculation. Let's say that you are removing a 150 hp Lycoming engine from your experimental 4 passenger Traveler airplane, and replacing it with a 220 hp Buick V-6 engine. The Lycoming, with all the baffles, magnetos, starter, alternator, prop spacer, and full of oil, weighs 345 pounds. The engine is located 29 inches forward of the firewall. The Buick V-6 weighs 390 pounds, and is located 19 inches forward of the firewall. The Buick V-6 engine is closer to the firewall because its 50 pound PSRU gearbox is 12 inches long and is ahead of the Buick engine. Next, we will do our calculations to see how the heavier engine affects weight and balance of the Traveler airplane. To make the solution simple, remember the formula for calculating weight and balance: "WEIGHT TIMES ARM EQUALS MOMENT". So, we have a 345 lb. Lycoming engine @ 29 inches forward of the firewall: therefore, we multiply 345 lb. x 29 in. = 10,005 in. lbs.. Next, we multiply the Buick V-6 engine wt. of 390 lbs x its 19 in. forward of the firewall = 7,410 in. lbs. This means that the Buick V-6 engine, although heavier, must be moved forward to properly balance the airplane.

DIVIDE MOMENT

To see where the engine should be mounted, take the moment arm (wt.) of the Lycoming engine, which we calculated @ 10,005 in. lbs., and divide the weight of the Buick V-6 into that figure (10,005 divided by 390 = 25.65 in. [ahead of the fire wall]). That is a simplistic method because it does not account for the total added weight of 45 extra pounds that is also distributed aft of the datum line. For further math to correctly locate your auto engine in your airplane,

Consult the new FAA publication AC-43,13B, "Aircraft Standards for Repair".

Exhaust Systems

◆◆◆◆◆◆◆◆◆◆◆◆◆◆◆◆◆◆◆◆◆◆◆◆◆◆◆◆◆◆◆◆◆◆◆

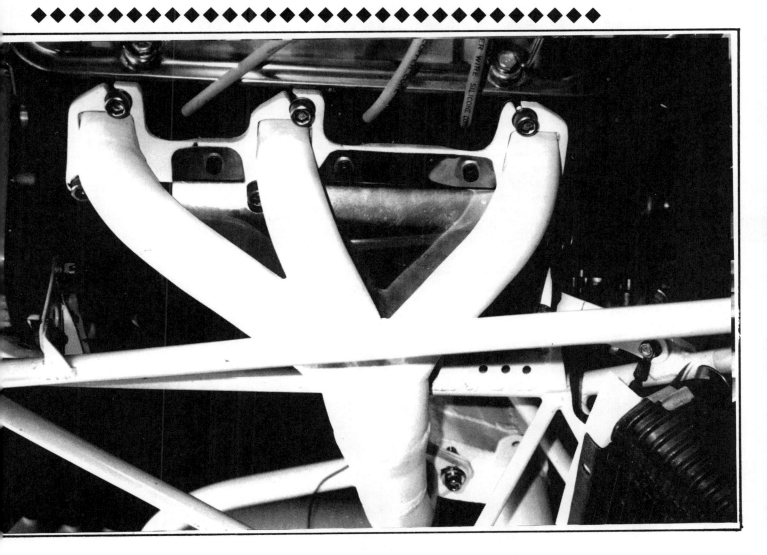

This ceramic coated exhaust pipe is the left side pipe on the author's Buick V-6 engine converted for aircraft use. The author simply purchased a race car header set that had been dyno tested and then modified the header to fit through the engine mount tubes. A "Flowmaster" type muffler is used also. Notice the stainless steel heat shield that protects the rubber mount bushings from exhaust heat.

TUNED EXHAUST

There is an easy 5 to 8% more horsepower available from most auto engines when you incorporate a race car type tuned exhaust, compared to a basic, restrictive exhaust pipe or a short stack type exhaust.

In addition to giving you a few extra horsepower from your engine, tuned exhaust systems also quiet the exhaust sound somewhat, and they also provide for a collector pipe that can be attached to a race car type muffler for even more sound suppression.

If you take a few extra hours to design and build a tuned exhaust system, it will pay off in the future by way of less heat in the engine compartment. Building a pair of racing exhaust headers takes 8 to 16 hours.

47

EXHAUST COATINGS

The exhaust systems pictured in this chapter are spray painted with VHT(TM) ceramic, heat proof coating. This spray paint, when properly cured in an oven at 400 degrees F., and then cured on the engine when the engine is running, will both insulate the exhaust heat, and protect the exhaust system from rust and heat deterioration.

Spray can VHT (TM) paint is available at most auto supply retail outlets, and is available in many colors, including white. Expect to pay about $10.00 to $15.00 for enough to completely coat one set of exhaust headers and muffler. The nice thing about VHT (TM) is that you can apply it at home or in your workshop.

Other brands of exhaust coatings work even better than VHT(TM), but they cannot be applied at home. They must be applied in a controlled condition factory setting. These high quality coatings insulate the heat into the exhaust pipes and out of the engine compartment. And they cost a LOT more, like $100 to $150 to coat one set of exhaust headers.

This "Flowmaster" type muffler weighs 7 1/2 lbs. and it quiets the exhaust to the level where the Buick V-6 auto engine can be operated in a residential neighborhood without any noise complaints.

However, they really work. Even after running the engine for many hours at full power, the coating remains as bright and smooth as the day it was applied. The exhaust system must be completely welded, fitted to the engine, and it should be test run to be sure there will be no further changes. Then it may be removed from the engine and shipped off to one of the off-site coating company locations. Check chapter 19 for the names and addresses of exhaust coating companies.

JET HOT COATINGS

These coatings are superb for making your exhaust system last a long time and for keeping the engine compartment cooler. Accelerated tests show zero deterioration when the coating is used to protect stainless steel exhausts, and negligible deterioration when used to coat mild steel exhausts. Once your exhaust system begins to rust and flake, it can still be coated with **JET HOT TM**.

Reiner Hoffman of Stratus, Inc., Seattle, Washington, uses an automotive type muffler to quiet his Subaru Legacy engine so well that flying this airplane is almost as quiet as a sports car driving down the freeway.

Photo: Reiner Hoffman

STAINLESS STEEL

For many years, the standard practice in certified aircraft engines has been to make the exhaust system out of stainless steel to prevent rust. Stainless steel will not rust, but it will corrode, flake off and crack. I doubt that any Piper, Cessna, Mooney, Beechcraft, or Grumman airplane ever flew its exhaust system more than 500 to 1,000 hours without having to undergo expensive repairs to its stainless steel exhaust system.

You can significantly improve the durability of a stainless steel exhaust system by coating it inside and out with one of the commercial insulating coatings such as Jetcoat(TM). But do not think that you absolutely must make your auto engine exhaust system out of stainless steel. Mild steel exhausts, coated with ceramic or Jetcoat(TM) will work very well.

This Mazda 13B rotary engine in an RV-4 airplane has a very compact exhaust system and a muffler that is absolutely mandatory for rotary engines. Photo: Tracy Crook

MILD STEEL EXHAUSTS

Say that you are installing a small-block Chevrolet V-8 into your Lancair IV experimental airplane. The cheapest, most effective, and easiest way to fabricate an exhaust system for your airplane is to purchase a set of racing headers for $100 to $150, and slightly modify them to fit your airplane. If they interfere with the already fabricated engine mount, simply cut, trim, and modify the store-bought headers to fit.

Doing the ready-made header installation will be fast and fun. You can even oxy-acetylene (gas) weld the mild steel exhausts to fit and work well. And of course after you have fitted the exhaust system, remove it and coat it with high-temperature ceramic paint or send it off for professional heat proof coating.

Mild steel exhaust systems can last for many years and many hours of flying if you don't allow them to rust excessively. Say that you have operated your airplane for 25, 50 or 100 hours, and you are sure the exhaust system is going to work as-is, then you can remove it, sand blast it to remove the rust, and coat it with **VHT** ceramic paint, or send it off for **JET HOT** coating.

The exhaust pipe system on the Legend is very simple. Each of the 8 stacks is nearly the same length and shape.

WELDING STAINLESS STEEL HEADERS

Both airplanes shown on this page have professionally welded stainless steel headers. Even though it is possible to gas (oxy-acetylene) weld stainless steel, don't do it. The high temperatures associated with aircraft exhaust system operating, will adversely affect the gas welds.

The best way to build headers like you see on this page, is to T.I.G. weld them. M.I.G. or wire feed welding works better than gas welding, but it is not as clean and strong as T.I.G. welding.

Another important trick in welding stainless steel headers, is to back-gas purge the inside of the exhaust pipe with argon gas before, during, and for several seconds after the weld is completed. Back-gas purging the exhaust pipe will prevent the stainless steel from crystallizing as it is welded.

If you allow stainless steel to crystallize, the "sugar", crystallization will overheat when the engine is

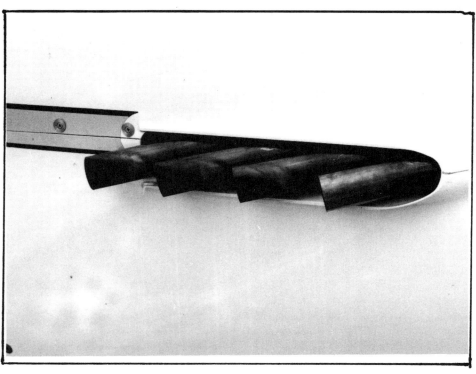

Jeff Ackland's Legend V-8 uses short, individual exhaust stacks for sound and visual effects.

operated, and cracks will from at the welds. Many small shops are equipped with T.I.G. welders and T.I.G. is easy to do. You just need an extra bottle of argon gas and an extra flow regulator when welding stainless steel. With a good T.I.G. welding set-up, you can also weld Titanium and aluminum.

TIG & MIG WELDING

These terms are often confusing to part-time welders. T.I.G. means Tungsten Inert Gas, which most people still call Heli-arc™. Except that Heli-arc™ is a trade name of a particular manufacturer, ESAB. It is like calling a refrigerator, a "Frigidare"™. Anyway, TIG welding is a very clean, precise process that gives you the control to weld almost all weldable metals, including stainless steel and also aluminum.

A good quality TIG welder sells for $1600 to $4000. For more information about TIG welding check out the books, "Welder's Handbook", published by H.P. Books, and "Performance Welding" published by Motorbooks International. Both of these books are also written by Richard Finch.

MIG welding is often called wire-feed welding. The initials stand for Metallic Inert Gas. The process works by feeding thin wire into the weld to make the arc.

Wanting a quiet but sporty sound, George Morse built this stainless steel header system and exited the dual exhaust out below the nose cowling, just like an aircraft exhaust system.

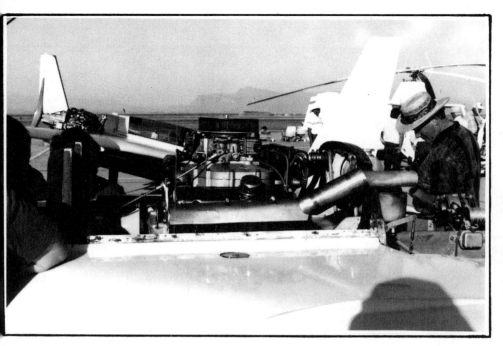

Because Shirl Dickey's E-Racer engine is mounted completely inside the rear-mount cowling, he has added heat shields to the entire exhaust system to prevent overheating the fiberglass structure of the airplane. Exhaust heat is often 1,000 to 1,400 degrees F.

HEADERS VS. BLUNT PIPES

Two parts of an auto (or aircraft) engine are very sensitive to smooth gas flow. The intake system will flow more cool air and fuel if the bends and seams in the intake manifold are smooth and gentle.

And equally so, the exhaust system will allow the hot combustion gasses to exit the engine with much less back pressure if the individual exhaust pipes have smooth, gentle bends. In hot rod days, we could develop up to 15% more power from our street rod engines if we put a set of exhaust headers on our cars, in place of the stock, cast iron exhaust systems.

Converted auto engines in airplanes that are only getting 150 hp to 160 hp from a 265 cubic inch engine, are choking the engine with restrictive exhaust systems. By merely changing to racing header exhausts, the engines can breathe better and let the engine produce more power at the propeller.

EXHAUST HEAT

Remember that the e.g.t. (exhaust gas temperature) gage on the instrument panel can show **1,100°F** to **1,350°F** when the craft is in flight. Providing good heat shielding for wiring, fuel lines, rubber parts and even the oil pan, is very important. Shield your exhaust.

There are a number of products on the market that will effectively shield exhaust heat away from components such as water lines, fuel lines, oil lines, your engine's oil pan, and the fiberglass nose cowling. One of these products is NOMEX cloth, covered on one side by an aluminum foil material. In most heat situations, a reflective material plus a very small amount of fresh air flow will effectively cool the hot spots. Also consider the effects of radiated heat soak after the airplane is parked on a hot ramp and the engine is shut down after flight. You may want to incorporate heat flaps in the top of your engine cowling.

If exhaust heat appears to be a problem in your engine installation, you can purchase Tempil™ heat indicating tape to record the highest in-flight heat reflected into your engine compartment. The tape changes color to show the highest heat attained.

Considering that this water cooled OX-5 V-8 engine was built in the late 1920's, this race car header exhaust system is very efficient. Note that it is gas welded mild steel, painted with aluminum paint to prevent rust.

HEAT BAFFLES

We all thought, way back in the 1960's and 1980's, that a side benefit of flying water cooled auto engines, was that we didn't need all that baffling and air seals in the engine compartment. But that is really not so.

In the engine picture on this page, you can see how effective it is to baffle the cool incoming air that flows through the radiator, and to shield the hot, 1,100°F to 1,350°F exhaust pipe heat away from the radiator and the rest of the engine.

Dr. Bill Harrold, Tom Jones, Jess Meyers and others in Las Vegas, Nevada, have developed an excellent method for cooling a V-6 Chevy engine, and for keeping the exhaust heat separate. This system works well in this Chevy V-6 powered RV-6A airplane, but it will also work in Ford V-8, Chrysler V-6, and Chevrolet V-8 powered Lancair IV and other similar size airplanes.

There will likely be improvements to this excellent design, because that is how we have developed airplanes since the early 1910's. We learn by experimenting. And experimenting (Experimental Aircraft Association) is what the movement is all about.

If you want a successful cooling system and effective exhaust system for your airplane, copy this very effective design.

HEAT SHIELDING

Not visible in the photo on this page, is a thin layer of cloth-type heat shielding (described on page 51) that prevents exhaust pipe heat from burning the fiberglass nose cowl material. Notice the closeness of the exhaust pipe to the fiberglass AND the decal on the outside of the cowl. NO scorching exists there.

Barely visible between the RV-6A nose cowl and the V-6 Chevy engine baffle are two of the three header-type exhaust tubes; each cylinder uses an EGT probe to measure individual cylinder temperature.   Photo: Jess Meyers, RV-6A

Ignition Systems

◆◆◆

Jess Meyers uses two ignition coils that are operated individually, to provide a redundant ignition system for his 4.3 Chevrolet V-6 engine converted for aircraft use. Here you see the two coils and the automatic coil splitter that fires single spark plugs.

DUAL IGNITION

The third question that people ask when they are wondering about auto engines in airplanes, is, "What do you do for dual ignition?" (The first and second questions concern weight and water cooling.)

The answer to this third most asked question is, "You can have dual ignition if you want it, but the FAA does not require it." The FAA says you can actually certify a spark ignition, piston engine with single ignition if you can prove that your single ignition system is equally

as dependable as a dual ignition, magneto operated system!

One of the most troublesome systems on a "modern" certified aircraft engine is its dual magneto, dual spark plug ignition system. Every person who has a private pilot's license or above, has at one time or another, done a pre-flight magneto check and has had to abort the flight because of an excessive mag drop (rpm drop). ANY CURRENT AUTOMOTIVE IGNITION IS MORE DEPENDABLE than any certified aircraft engine ignition system.

HOWEVER...if you were taught to do mag checks before each flight, and you can't bring yourself to break that habit, there are ways you can have dual ignition on your converted auto engine. There are several different ways you can incorporate dual, triple, quadruple, or even more ignition systems, if you feel that you simply must do that. And I agree that an airplane engine that suddenly quits running can scare you right out of your pilot's seat! Like when the single drive gear on your dual magneto CERTIFIED aircraft engine fails!

JESS MEYER'S SYSTEM

In one of the most simple dual ignition systems to date, Jess Meyers adds a second set of ignition points to his Mallory(TM) distributor, and he wires each set of points to a toggle switch on the instrument panel, then to a separate condenser, and then to the plus side of a separate ignition coil.

Then each coil is wired to a special coil wire splitter that decides electrically (by a diode) which coil is firing. Then the coil splitter feeds the high voltage spark to the distributor cap and the rotor inside the distributor cap feeds the spark to the proper spark plug.

Automotive spark plugs last 100,000 miles or more and they cost $2.00 apiece, so there is no earthly (or heavenly) need for dual spark plugs in Jess Meyers system.

This dual ignition system will also work with L.E.D. or magnetic impulse distributors, by simply adding a second L.E.D. window and pick-up or a second magnetic spark pickup to your aftermarket distributor.

FACTORY SYSTEMS

If you presently drive a .GM auto with a V-6 engine, you are depending on its factory supplied **TRIPLE** coil ignition in which each pair of cylinders is fired by a separate coil. If you drive a GM V-8 auto, you have **FOUR** separate ignition coils firing your eight spark plugs. And your engine spark timing is automatically adjusted for the best performance as you start the engine, get onto the freeway, and drive to the ski lodge at 12,000 feet elevation, all the while pulling your heavy camping trailer.

The new automotive ignition systems are totally electronic, with NO moving parts to break or wear out, and they are no more prone to failure by lightning strikes than the fine-wire windings in your certified aircraft magnetos are.

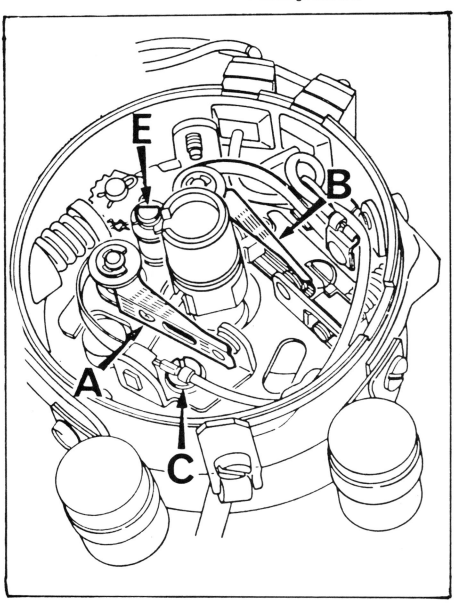

An extra set of ignition points can be installed in most distributors to provide a redundant ignition system. "A" is the primary set of points, "B" is the new, secondary set of points. The two condensers are mounted on the outside of the distributor housing.

Here you see THREE separate ignition coils for a triple ignition system (not dual) on a supercharged Buick V-6, 240 hp engine. This is Buick factory stock. This is a factory cut-away engine for marketing display purposes.

JEFF ACKLAND SYSTEM

It appears that the Legend V-8 engine employs a dual coil, dual points, coil splittler system much like the Jess Meyers dual ignition system. One thing to avoid is bundling the spark plug wires together rather than to space them apart.

When spark plug wires are bundled or tied together, or even just touch each other, they can cross fire and cause severe piston damage due to detonation/pre-ignition. Spark plug wires spacers cost less than $5 for a full set. Buy a set and use them to prevent holes in pistons or even a burned out ignition system.

MSD aftermarket distributors can be easily converted from single ignition to a second, back-up ignition system. Other brands, such as Malory and Accel can also be converted to dual spark pick-up.

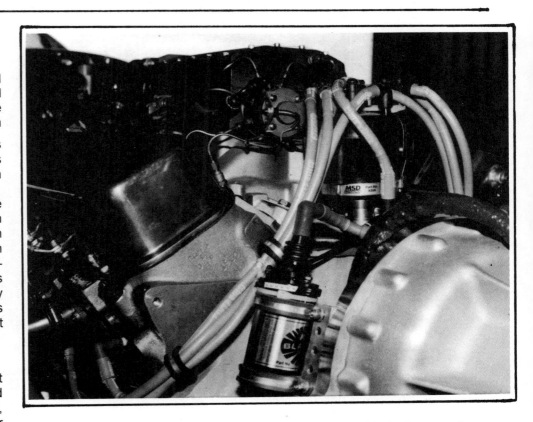

Jeff Ackland's Legend big-block V-8 engine is ignited by a MSD distributor and two separate coils, similar to Jess Meyer's ignition system. One coil is shown here, and the second coil is on the opposite side of the engine, firing a single spark plug in each cylinder.

ELWYN JOHNSON SYSTEM

In this Chevy V-6 ignition system, Elwyn Johnson has incorporated two MSD electronic, race car type ignition systems that are capable of firing the spark plugs at up to 10,000 rpm. Each MSD ignition system can be electrically switched to provide separate and redundant ignition to fire the spark plugs.

If the system should short out, or stop doing its job, the other MSD system will continue to operate. In this system there is only one set of spark plugs, but each $2 spark plug should last 100,000 miles or more.

Elwyn Johnson has installed dual and redundant MSD electronic ignition systems on this 4.3 Chevrolet V-6 engine in his Cessna 172. You can see the two MSD modules on the firewall, above the pilot's left foot area. A cooler location would be better.

Photo: Elwyn Johnson

CRANK TRIGGER SYSTEM

Most race cars now use a very accurate electronic hall-effect sensor that picks up evenly-spaced pulses from a wheel that is externally bolted to the engine crankshaft, and more and more production automobiles actually cast a trigger wheel into the engine crankshaft.

Then a special magnetic pickup is screwed into the engine block to sense the pulses from the factory designed trigger wheel.

With an add-on trigger wheel, it would be easy to add one, two, or three more electronic pickups, and have as many separate ignition systems as your fears dictate.

Crank trigger ignition and electronic fuel injection systems will eventually replace distributors because a stock distributor is driven by the timing chain on the cam, then a gear, which causes lots of spark variations.

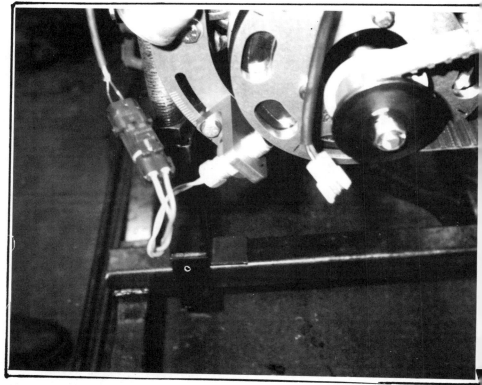

Ken Duttweiler mounts this crank trigger electronic pickup on the front pulley of his Buick V-6, Chevrolet V-6, and Chevrolet V-8 engines. It provides variable spark timing through an Australian made electronic ignition and fuel injection system.

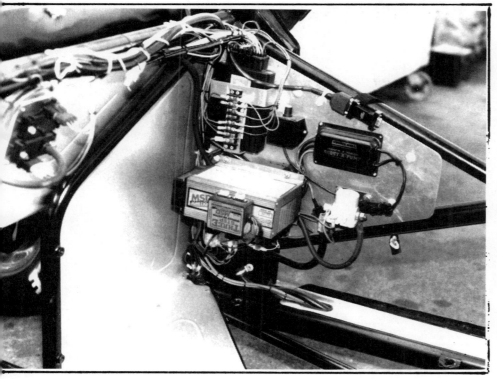

LAP TOP DOWN LOADING

As with most of the after-market electronic ignition and fuel injection systems, there are ports in the control boxes that provide for instant down-loading of engine performance data into a lap-top or palm-held computer.

It is even possible to monitor all the functions of engine performance while you fly, by continuously reading the outputs of timing, fuel burn, mixture, oxygen sensing, manifold pressure, throttle position, rpm, and etc. right on the computer screen, and if a fault occurs, the computer can tell you what it is and give you a chance to correct the fault in flight.

As the cost of lap-top computers continues to drop, more and more pilots will want to use this fabulous feature.

There are more electronic "brains" in this picture than can be explained in a single book chapter, but Den Duttweiler can plug in to this MSD box and can download his Buick V-6 operating data into his lap top computer in a matter of seconds!

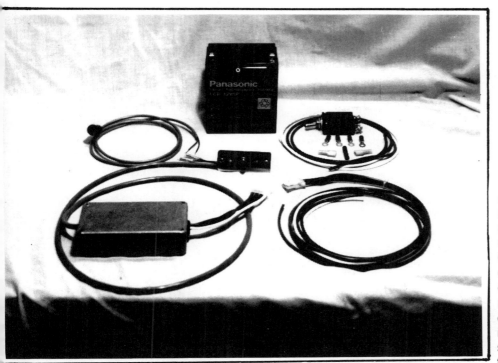

OLSEN TECHNOLOGIES

If you plan to fly dual, electronic, automotive ignition on your airplane, you should check out the ignition system condition analyzer invented by Olsen Technologies. This system can let you know if you have a "sick" ignition system and it will provide a "limp home" mode.

HEI SYSTEM

For instant starts in hot or cold weather, the GM High Energy Ignition system cannot be beat. It is a very hot, dependable system for engines that do not operate above 6,000 rpm (all our converted auto engines in airplanes), and if you change the $5.00 rotor every 25,000 miles, it will last longer than your engine will. But you cannot convert it to make dual ignition unless you stack one HEI distributor on top of the stock one. And that is not a practical thing to do.

Olsen Technologies makes this electronic back-up system for battery operated ignition systems. If you are not using an antiquated magneto, you should investigate this safety ignition accessory.

Just above the starter in this picture, you can see a Buick -invented knock sensor. It looks like an oil pressure sending unit. You should try to install at least one or two of these on your engine to detect detonation knocking.

KNOCK SENSORS

Buick developed knock sensors for the Turbocharged V-6 Grand National (Regal Coupe) in 1980. The knock sensor is threaded to screw directly into the engine block. When the sensor "hears" a knock caused by poor fuel, lean mixture, or excessively fast timing, it electronically retards the timing to stop the knock. If the knock persists, a fault code light is displayed on the instrument panel, telling the driver to "check engine".

You certainly should incorporate knock sensors into your electronic ignition on your airplane.

Even if you install a knock sensor (or two) without wiring it in to your electronic ignition, it can still be wired to a knock warning light to tell you that your ignition timing is too fast, or that your fuel is too low. in octane.

AUTO FACTORY IGNITION SYSTEMS

The most dependable ignition systems you can have on a converted automobile engine is the one that the factory designed, tested and produced your auto engine with. In sheer numbers of hours of operation, a wild guess is that auto ignition systems have literally BILLIONS more hours of dependable operation than certified dual magnetos do.

The most likely in-flight failure will not be your factory ignition system. It will be your battery or your alternator. Rebuilding and proof-testing your alternator on a regular basis, say once every 5 years, will do more to provide dependable sparks to fire your engine than modifying a factory ignition system.

Dual batteries with a fool-proof charging and switching system is your best insurance. Check into RV or motorhome charging and switching systems if you fear failure.

In my own experience, dead batteries and failed charging systems (alternators) are the number one cause of ignition failure, not the ignition system itself.

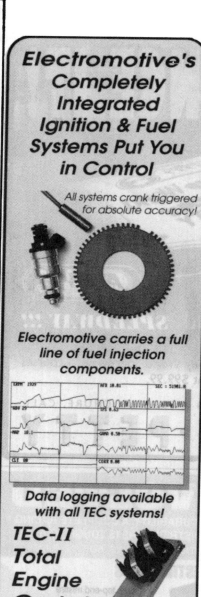

Fuel Injection & Carburetors

◆◆◆◆◆◆◆◆◆◆◆◆◆◆◆◆◆◆◆◆◆◆◆◆◆◆◆◆◆

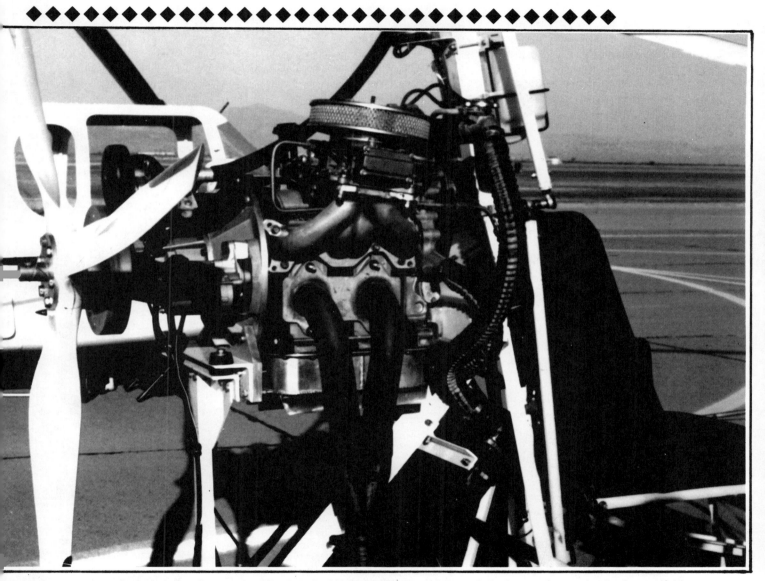

Easily the least complicated of all carburetor modifications is this 2-barrell Holly carburetor bolted to an after-market intake manifold, as is used on this Mazda rotary powered gyro copter. Changing from factory stock takes 2 or 3 hours.

AUTO ENGINE G.P.H.

An established scientific fact is that an air cooled engine needs a richer fuel mixture in order to produce 200 horsepower than a water cooled engine needs to produce the same horsepower.

The reason for this requirement for extra fuel is that an air cooled engine needs the cooling effect of a comparatively rich fuel mixture in order to cool the engine and prevent the pistons from seizing in the cylinders. This extra rich mixture in a current design certified aircraft engine, can amount to as much as 20% more fuel used each hour. This means that an airplane powered by a water cooled engine, can carry less fuel and go just as far.

VOYAGER FLIGHT

A prime example of the use of water cooling in an aircraft to extend the range of the aircraft was the use of a water cooled engine in the rear of Dick Rutan and Jenna Yeager's 'round the world record setting Voyager aircraft. The engineers planning that flight realized that the airplane could not have flown the record setting flight powered by an air cooled engine because of the 15 to 20% higher fuel consumption of air cooled engines. They would have run out of fuel.

So they convinced Teledyne Continental Motors to add water cooling jackets to the pusher engine in the rear that would be run 100% of the time of the record setting flight. And that flight was made with a relatively inefficient 1930's design up-draft carburetor. Today's modern electronic fuel injection would have given the Voyager even better fuel economy. Using a water cooled engine contributed very, very significantly to that record setting flight success.

All (100%) of today's autos use electronic, O_2 sensing, barometric pressure, altitude sensing fuel injection systems, which are ideal for our converted engines in our aircraft.

Ken Duttweiler changed this Chevrolet V-8 engine to use electronic fuel injection, for more power, smoother operation, and better dependability. A very natural change is to go to electronic fuel injection rather than fight the age-old problem of trying to fit and tune a carburetor for best power at all altitudes, all temperatures, and all fuels.

This is the front view of the Ken Duttweiler Chevrolet V-8 shown in the photo above. Except for the fuel pressure regulator that sticks up above the throttle plenum chamber, the intake system is very low profile, a desirable factor in fitting everything in under the cowling.

At the present time, fuel injection is about 5 to 10 times more expensive than carburetors are. A new carburetor will cost $200 and the new aluminum intake manifold will cost another $150. A fuel injection system will cost $2,000 or more.

An option would be to adapt an electronic fuel injection system from a wrecked car for $500 or less.

With the Adjust-A-Jet from Percy's High Performance (573/346-4409), fuel is metered through a single externally adjustable needle and seat orifice, allowing each venturi to get what it demands, rather than to what it's restricted, therefore creating a closer air/fuel ratio venturi to venturi.

In the photo at the bottom of the page, you can see the GM type electronic fuel injection parts that have been adapted to this Buick engine, hot-rodder style. You can also see the 90° curved inlet for air, that is connected to a throttle body pirated from a car in a salvage yard.

Once you get past the fear of electronic devices in your engine compartment you should be able to convert your auto engine to fully automatic electronic fuel injection. Check out the 4th edition of the H.P. Book's "TURBOCHARGERS" for information about how to select and install electronic fuel injection systems.

Ken Duttweiler prepares to analyze the data that he just downloaded onto his lap top computer from his MOTEC fuel and engine management system on his drag race car. This entire process takes about 5 minutes. In an airplane, the lap top computer can give you real-time information about engine performance while you fly.

This 231 cubic inch Buick V-6 engine has been adapted to direct port fuel injection. Note the injector boss receptacles that have been welded to the intake manifold.

HOT ROD FUEL INJECTION

Steve Parkman of SWAG Aeromotive, Tucson, Arizona, has successfully adapted GM and Ford Fuel injection components to several experimental airplane engines. Quite likely, they could even adapt a certified aircraft engine to electronic fuel injection if they wanted to.

Here on this page, you see an air cooled VW engine that is sprouting a GM type fuel rail and two electronic fuel injection nozzles. These systems operate best when triggered by a magnet imbedded in an aluminum wheel bolted to the crankshaft.

The bottom picture shows how simple it is to fabricate a simple intake chamber with four ports that feed each of four cylinders. The Ford Topaz throttle body is simply

an air flow butterfly valve with an electronic throttle position p attached, larger, streamlined tube to each cylinder would add a fe horsepower.

This photo clearly shows the electronic fuel injection components that Al Haralson, of Tucson, Arizona, has incorporated in his air cooled Volkswagen engine conversion. If you can fuel inject a VW engine, you can easily fuel inject ANY brand of piston engine.

GM & FORD FUEL INJECTION

If you are considering electronic fuel injection, go to an auto salvage yard that allows the customer to "pull your own parts", and find a salvageable fuel injection system on a car with an engine of the approximate HORSEPOWER of your converted auto-to-aircraft engine. In other works, if you expect 175 horsepower from your Ford, Buick or Chevrolet V-6 engine, then you would look for a fuel injection system from a 4.3 liter Chevy V-6 engine.

Take a shop manual with you, and get ALL the parts off the engine and out of the car. For sure, you will need the computer and all the O_2 sensors, temperature sensors, and the wiring harness.

Steve Parkman, SWAG Aeromotive of Tucson, Arizona, fabricated this fuel injection air manifold for a GM Saturn engine that he has installed in a Vari-Eze. The throttle body is from a Ford Topaz auto.

This mechanical fuel injection system, made by Airflow Research, fits big block, 600+ hp engines. The Legend airplane uses one of these systems. Photo: Airflow Research

Airflow Research makes this Chevy V-8 small block mechanical fuel injection system for engines up to 450 horsepower. It is direct-port injected and has a very low profile. Photo: Airflow Research

Airflow Research also makes a mechanical fuel injection system for the Subaru Legacy engine, pictured here with a Ross PSRU, mounted in a Murphy Rebel kitplane.

MECHANICAL FUEL INJECTION SYSTEMS

If you have flown a Piper Arrow, a Ted Smith Aerostar, or other airplane with fuel injected engine, you have managed the fuel delivery by way of mechanical fuel injection, usually manufactured by Bendix. It works somewhat like the fuel delivery system on a port-injected diesel engine. Each intake port has its own fuel injection nozzle that either squirts a manually metered supply of fuel into the intake valve, or by a sequential fuel injection system, controlled by a rotary fuel delivery valve.

The continuous spray of metered fuel is the most common. It is similar to a vacuum operated carburetor system that constantly draws a 14.0 to 1 air and fuel mixture into the intake manifold when the pistons are moving. But the port fuel injection is more efficient because it does not wet the entire intake manifold, just the ports.

Mechanical fuel injection is far superior to a single throat, two throat or 4 throat carburetor because it atomizes the fuel better and because it puts the wet fuel exactly where it is needed, at the intake valve.

Airflow Research can provide mechanical fuel injection systems for any engine you would want to fly in your airplane, certified or automotive.

NOTE:

The captions on the top two photos were accidentally switched. The top unit is for a Chevy V-8 small block engine. (R.F.)

William Wynne, of Daytona Beach, Florida, has mounted a single motorcycle type side-draft carburetor on a fabricated "Y" type intake log, to operate his direct drive Corvair engine in a Casutt Racer airplane. With intake tubes this long, carburetor heat is a must.

Rich Dietrich, of Morrice, Michigan, has installed a Gates Poly-Chain drive in his Corvair engine, and uses a pair of Chevrolet Corvair carburetors to power his WW I replica biplane. This is a simple but effective solution to the carburation problem.

The highway sign reads: "Eisenhower Memorial Tunnel, Elevation 11,013 feet". The picture is blurred because we were traveling uphill at 75 mph, with a climb rate of 800 ft. min. in our V-6 Chevrolet auto. AND, we did not "lean out" the carburetor because most automotive fuel systems are automatic lean, barometric pressure adjusted mixture systems. Perfect for an airplane!

GOOD OLD CARBS

The two Corvair engines shown on this page use tried and proven single barrel carburetors. The engine in the top photo uses a single throat side draft carburetor, and the engine in the middle picture uses two $10 apiece Corvair single throat carburetors. Not much could be done to make the fuel systems in these two engines more simple or less expensive.

4 BARREL CARBS

Auto engine converters have found that 4 bbl carburetors with progressive secondary throttle opening, are not exactly good for airplane flying. That is because the secondary throttle valves usually open and close abruptly with almost no precise control by the pilot. This causes abrupt power changes in the engine, not what you want when trying to land a plane. Often, a 2 barrel, 500 CFM carburetor is much better on an airplane engine that a 4 barrel.

Most successful auto engine converters use the 350 CFM or the 500 CFM Holly 2 barrel carburetor with a manual leaning block adapted to the carburetor. GM Rochester 2 barrel carbs work well also. Cost of a used GM carburetor is $10 to $20.

LEANING PROVISIONS

One of the least necessary things to be concerned about, is providing manual leaning for your aircraft carburetor. We all drive our autos from sea level to mountain passes as high as 12,000 feet (see elevation of Eisenhower Pass, near Denver, Colorado) and we NEVER reach for a knob to lean out the fuel mixture in our autos. Leaning is just a habit we acquire when being taught how to properly fly AIR COOLED certified aircraft engines.

Radiator & Heater Designs

◆◆◆◆◆◆◆◆◆◆◆◆◆◆◆◆◆◆◆◆◆◆◆◆◆◆◆◆◆◆◆◆◆◆◆◆◆◆

This is the most efficient radiator in a kitplane so far. The unusually small radiator measures 29 3/4" wide, 8 1/4" tall, and only 2 3/4" thick. This chapter explains why it works so well.

RV-6A, CHEVY V-6

In the two photographs of the RV-6A Chevy V-6 installation in this chapter, you can see the custom made aluminum radiator mounted above and behind the engine. The radiator is only 40% as big as the radiator that is used in V-6 powered Chevy pickup trucks and vans. But the designers, Jess Meyers and Dr. Bill Harrold, have effectively baffled the engine heat away from the radiator, and have directed the incoming cool air to force all of it to flow through the radiator at flying speeds of 60 mph to 220 mph.

In effect, they have created a large plenum chamber above the engine to cause the air coming through the stock RV-6A nose cowl front inlets to be packed into the radiator fins for the most effective cooling seen on a liquid cooled aircraft to date. Other designs may eclipse this one in the future, but as this book goes to press, this cooling system design is state of the art.

Side benefits from this design are the ability to have short radiator hoses, no external plumbing outside the engine compartment, and a substantial weight savings because the cooling system is very compact and very tidy.

In this earlier version, Jess Meyers found that if he baffled engine heat away from the very small radiator, it would cool the engine very well, even at the 115 @F. runway temperature of Las Vegas, Nevada.

RADIATOR & HEATER DESIGN

The single most significant factor in designing a dependable auto engine installation in an airplane is the correct design of the cooling system. If you have the world's most highly developed V-6, V-8, V-10, or V-12 engine in your airplane and you can't keep it from overheating, you have wasted all your time and all of your money on the project. This chapter will tell you how to properly cool your engine.

CUBIC FEET OF AIR

A liquid cooled engine radiator is technically a heat exchanger. The engine and its water pump puts hot water into the radiator, and cool air passing through the radiator extracts the heat from the liquid and when it is returned to the engine, it is able to soak up the heat and the cycle continues. The radiator exchanges hot water into cooler water.

The clue to effective cooling of an auto engine in an airplane is the passing of a specific number of cubic feet of cool air per minute through the radiator.

There is a formula for calculating how much radiator area is needed to cool a specified horsepower engine at a specified speed. The engine will produce a specific number of BTU's of heat per minute at a given power output. A radiator opening of a specified number of square inches will pass a certain number of cubic feet of air through the inlet at a given speed in feet of forward movement per minute. But calculations on paper are often not fully representative of the actual operation of machinery. Even though you can calculate something on a computer or on paper, the proof is in the testing.

THE REAL WORLD

Every radiator design shown in this chapter works. There is no single absolute way to design a radiator and cooling system installation for your airplane. Take a close look at the radiator installations to see which ones suit your airplane the best. As we said in previous editions of this book, "If you want to win, copy the winner, not the loser." In the next few paragraphs, this chapter will explain the cooling systems that have been tried that failed. Then this chapter will suggest several cooling systems that work very well, and will explain why they work so well.

WHAT DOESN'T WORK

Previous designs that failed to cool the engine properly were radiators that were not designed to force air flow through the cooling fins, cooling system plumbing that allowed steam and air pockets to be

In Jess Meyers earlier RV-6A airplane, he used two GM air conditioner evaporators mounted in the nose of the airplane, and they cooled the Buick V-8 engine very adequately in the heat of the Nevada desert.

This Subaru powered airplane uses two GM air conditioner evaporators in the normal air inlets to cool the engine, plus a third GM evaporator under the prop to cool the engine oil.

This is what the GM (General Motors Harrison) air conditioner evaporator looks like as it comes out of a car or a truck. The units operate under very high pressure in the automobile air conditioning system (200 - 400 psi) and many of them split a seam and leak after 40,000 to 50,000 miles of use.

The leaks can be repaired by welding, and when these heat exchangers are used to cool an auto engine in an airplane, they never see more than 15 to 20 psi radiator pressure. Many air conditioning shops will sell you one-or even a dozen-for $5 apiece, because they are just scrap for auto air conditioning use.

Tracy Crook of Florida has used a pair of the auto air conditioner evaporators in his Mazda rotary powered RV-4 airplane. New evaporators cost about $200 from the dealer or the auto parts store.

trapped at high points in the system, radiators that attempted to change the rearward flow of air to 180 degrees or more, even 360 degrees in one specific instance.

Other things that did not work were radiators mounted where hot air off the engine passed through the radiator, defeating the heat exchange properties of the radiator.

AIR CONDITIONING EVAPORATORS

A very popular way to effectively cool auto engines in aircraft, is to adapt 2 or more auto air conditioning evaporators to take care of the heat exchange from the engine to the radiator and back to the engine.

AC evaporators are the part of the refrigerator system that fits inside the car or at the firewall to blow (or suck) air through to cool the car. The other heat exchanger, the freon condenser, is located ahead of the auto engine radiator, to extract the heat from the system.

A/C evaporators make very good radiators that can be positioned almost anywhere in the airplane, even in a belly scoop near the tail, or in under-wing scoops like a British Spitfire. But be extra careful when welding radiator fittings on the freon lines. The oil in the evaporator can explode with deadly results. Always purge the core with Argon or even with water to prevent gas explosions when welding.

CUSTOM RADIATORS

If your airplane building budget will stand $500 to $750 for a custom fitted aluminum radiator, then you will need to contact Griffin Radiators or one of the other radiator building companies listed in Chapter 19.

Take a close look at several of the successful radiator designs shown in this book. And like we have been telling you throughout this book, "copy the winner if you want to be a winner." Remind yourself that trying to be different can be both expensive and embarrassing as well as being quite disappointing when trying to design a radical new cooling system.

Stick to the laws of physics when you are designing your cooling system. Remember that air rises and water falls in a tank or a cooling system. Remember that air does not like to turn corners at 150 mph to 250 mph or faster. But also remember that the number of molecules of air flowing through a radiator determines how effective the radiator will be. Faster airplanes do not require as much radiator area as slower airplanes do.

SUITABLE RADIATOR LOCATIONS

Take a look at all the successful auto engine radiator locations shown in this book. They all work acceptably or even outstandingly. V-6 and V-8 engines will usually have room for a radiator just above the water pump, behind the engine.

An ideal cooling system is one where the radiator and the coolant are above the level of the engine, especially above the WATER PUMP. But certain engine and airplane designs do not have room for a radiator at this location. If your engine installation will not fit a radiator there, you can place it almost anywhere so long as you do not have high places in the water lines. High places will trap air, and air will not let water flow.

George Morse, of Redding, California, mounts one GM air conditioner evaporator under each wing, in a Spitfire-type scoop as shown here on his Jaguar V-8 custom built sport plane. Cooling has always been very adequate with this system, even while taxiing in hot weather.

Griffin Radiators in Georgia can custom make radiators such as this one shown here. You furnish a 3-view drawing of the radiator that you need, and they will build it for $400 to $600, depending on the size and shape you need.

Talk to your own local radiator shop. They may be able to TIG weld a custom size radiator for you for even less money.

There is much to be learned by carefully studying this picture of John Harlow Jr.'s Chevy L.T-1 Corvette installation in his Lancair ES 4 place airplane. First, you will notice the bolt-on race car exhaust system that fits perfectly. Next, you will notice the factory stock LT-1 Corvette engine with the Northwest Aero belt drive installed. Then you will also notice the factory stock size Chevrolet Camaro radiator mounted behind the engine. John Harlow, Jr.'s flight tests and ground runs clearly pointed out the fact that the radiator was picking up heat from the engine and especially from the exhaust system.

The solution to the reverse heat exchange in the radiator, was to shield the exhaust system and build cool air ducts to the radiator, an extra radiator was also added to the bottom of the Lancair ES.   Photo: John Harlow, Jr.; Tucson, Arizona

RADIATOR SCIENCE

A full-size Ford station wagon, pulling a large camping trailer up a long hill in the Desert Southwest on a hot August day, must have a suitably designed radiator system, or the V-8 engine will overheat and destroy itself.

Auto factory engineers and designers must design the cooling systems for the absolute worst operating conditions, and they must make them work, or the auto buying public will sue them, stop buying their cars, and soon put them out of business.

So, as one of my early engineering managers once told me: "The factories have spent millions of dollars to perfect this ___ (brand XYZ) car, so learn from them..." This means, if you want to save yourself a lot of trouble and expense, go look at the radiator on your family car:

1. It is out in front of the engine where it sees only cool fresh air
2. It is as high in the engine compartment as possible to assure that the coolant will always flow back into the engine
3. It utilizes a coolant recovery tank to allow for coolant expansion and contraction
4. It incorporates a thermostat to

This picture shows the bottom mount P-51 type radiator added to help cool the engine. Even this additional radiator is picking up some heat from the two exhausts that exit at the lower part of the firewall. John Harlow, Jr. says that one small radiator above and fully baffled from the engine might resolve the heating problems. Photo: John Harlow, Jr.

keep the engine at an optimum operating temperature of 180°F to 190°F.

5. And it is NEVER positioned where hot air from the engine and the exhaust system can flow through the cooling coils.

Unfortunately, there is not much chance that we can place the radiator at the front of the engine, but with thought and careful planning, we can direct cool air to the location where a radiator can be placed.

COMPOSITE AIRPLANES

With composite airplanes such as the Lancair ES shown on this page, you can't cut into the fuselage or wing skins to install a radiator as you can safely do with a tubular steel fuselage or a modular aluminum wing design.

The chore of developing a suitable cooling system is not as easy with a composite structure. But it CAN be done. Consider mounting external belly scoops or external spitfire type scoops under each wing. The Lancair IV and Lancair ES airframes can also utilize

a Jess Meyers type-half-radiator above and behind the V-8 engine, with suitable heat baffling, of course.

But as with most successful designs, the first attempts at a design are not always 100% successful. Unfortunately, we need a big corporation research and development budget to be able to afford trying out many types of cooling systems designs.

HOT AIR IN - HOT WATER OUT

It makes common sense that if hot air off the engine is allowed to pass through the cooling fins of the radiator, there will be little or no cooling accomplished by the radiator. If your engine coolant temperature at the outlet is 220 degrees F., and the air coming off the engine and the exhaust system is 400 degrees F. or more, you will actually <u>HEAT</u> the radiator coolant and boil it in place of cooling it.

By all means, avoid passing hot air through the radiator. If your installation requires that you mount the radiator at the engine firewall area on a tractor (NOT pusher) engine installation, the radiator must be ducted to prevent heated engine air from affecting the cooling process.

COOLING SYSTEMS THAT WORK

Several very successful auto engine conversion people have effectively proven that an airplane flying at 150 to 250 miles per hour dues not need as much radiator capacity as a van or a motorhome

pulling a boat trailer across the desert at 60 miles per hour on a 120 degree F. hot day.

In this later photo of John Harlow, Jr.'s., Chevy V-8 Lancair ES, you can see how John has added large diameter aluminum ducting to bring cool air to the radiator. Before these ducts were added, all the air going through the radiator was heated by radiated engine heat and by 1,400°F. exhaust pipe heat. Not a good way to cool a radiator from 200°F. down to 150°F.

The engine in both modes of transportation may be operating at 80% power, but the airplane is passing three to five times as many cubic feet of air per minute through the radiator as the van or motorhome is at the same load settings. For all practical purposes, the number of cubic feet of air passing through the fins of a radiator determines how many BTU's of heat can be extracted from the radiator each minute.

MULTIPLE RADIATORS

In certain tightly cowled engine installations, it will be necessary to spread out the cooling chore of the radiator into several separate but effective locations.

For instance, in the Buick V-6 engine installation, the H.E.I. distributor is at the rear of the engine where it interferes with a full width above-engine radiator. The solution here is to have one radiator to the right side of the distributor, a connecting hose to the outlet of the small radiator, and then a second, larger radiator to the left of the H.E.I. distributor. The outlet hose on the radiator then returns to the engine water pump inlet.

In other cases, you may want to save money and use a second-hand air conditioner evaporator core. Experience has proven that two of these cores will effectively cool a 300 cubic inch V-8 engine in flight and on the ground. And again, if you find that you have high spots in the plumbing system that will trap air, re-route the hoses and the plumbing or at least install a bleed valve to bleed out the air when filling the cooling system.

HOSE SIZES

One thing that is not parallel with auto engines in aircraft, is the required size of the coolant hose. Most auto engines use 1 1/2 " to 2" radiator hoses. That is because the driver may be stuck in traffic in Phoenix, Arizona on a 115° F day,

Engineair of Daytona Beach, Florida, has installed a turbocharged aluminum small block V-8 in their Lancair IV-4 airplane, and they found it necessary to place 4 separate radiators in suitable spots in the engine compartment to properly cool the engine. This airplane also features refrigerated air conditioning which further complicates the very tight engine installation. But everything works very well. The obvious object in the picture above is the electric fan that helps cool the radiators during ground operation.

pulling a trailer and with his air conditioner going full blast.

Or the auto driver may be pulling a boat trailer up the long hill west of Needles, California, on a summer day. That particular hill is a true test of cooling capacity of any vehicle. In either case, the engine may not be pumping very much water through the radiator, and that is why auto radiator hoses are so big in diameter.

On the other hand, an aircraft climbing at 90% power will also be moving through the air at 90 mph, meaning that the radiator in the airplane is not having to work as hard to cool the V-6 or V-8 engine.

COOLANT

Remember that antifreeze can be flammable if it sprays out onto a hot exhaust pipe. Therefore, it is very important to safety wire all hose clamps, and if possible, add a second hose clamp to all hose connections. Don't allow coolant leaks. Most engine converters use a 50/50 mix of distilled water and ethylene glycol antifreeze.

A N FITTINGS

If you take a few hours longer and a few dollars more, you can effectively eliminate all coolant leaks by using braided hose and AN flare fittings in your auto engine installation.

Almost all aftermarket automotive suppliers offer AN coolant fittings. You will need a AN-14 to AN-16 (3/4" to 1") fitting to replace the thermostat outlet fitting and you will need an equal size AN fitting on the water pump inlet. And of course, the fittings in the radiator (s) will have to be the same size. The Griffin radiator shown on page 68 has an AN-6 fitting welded to the top tank. Again, several years of successful engine installations has shown that most V-6, V-8, and rotary auto engines will cool adequately with 3/4" ID to 1" ID coolant lines. You do not need the larger automotive hoses.

NOTE: At an Oshkosh Forum this past year, I heard one engine converter recommend water-only in airplanes. This should be **AVOIDED** because of the rust problem due to water only, and the likelihood of your radiator (not the engine) freezing at altitudes of 10,000 ft. and above, even in summertime.

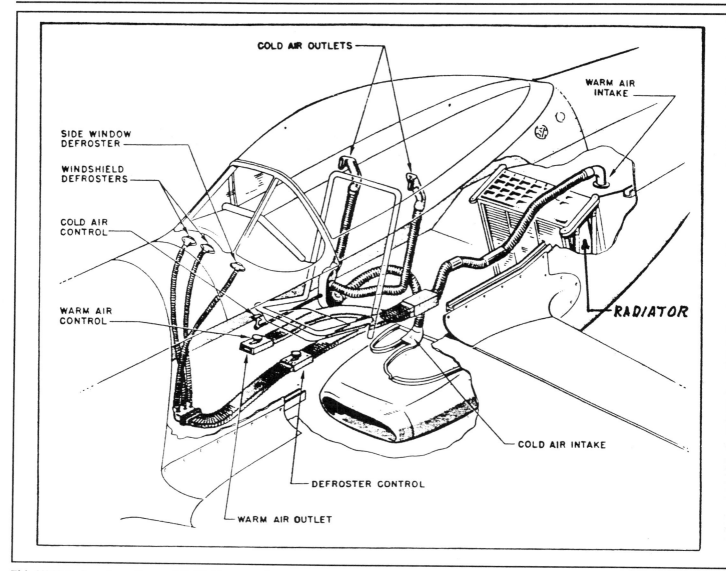

COLD AIR OUTLETS

WARM AIR INTAKE

SIDE WINDOW DEFROSTER

WINDSHIELD DEFROSTERS

COLD AIR CONTROL

WARM AIR CONTROL

RADIATOR

COLD AIR INTAKE

DEFROSTER CONTROL

WARM AIR OUTLET

This WW-II drawing of a real P-51 Mustang heating and ventilation system should give the experimental aircraft builder some good ideas about using exhausted radiator heat to heat the cabin of his water-cooled auto engine conversion. In the P-51 Mustang, the cabin air is heated from a scoop placed in the outlet of the engine radiator air. DWG courtesy U.S. Army Air Force, 1944.

COOLING THE ENGINE

The classic of water-cooled aircraft engines is the P-51 Mustang. Routing heated water from the engine to a belly-mounted radiator was ·the solution for cooling this WW-II V-12 engine. This chapter explains how it works.

HEATING THE COCKPIT

As long as you have heated air coming off the radiator in a P-51 mustang-type cooling system, you might as well bleed some of it off to heat the pilot and passengers with. This page is a good example of one way to heat the cockpit.

Another way to heat the cockpit and defrost the windshield would be to incorporate a simple heater core radiator from a typical small-car heater. The cost is low and the results are very satisfactory.

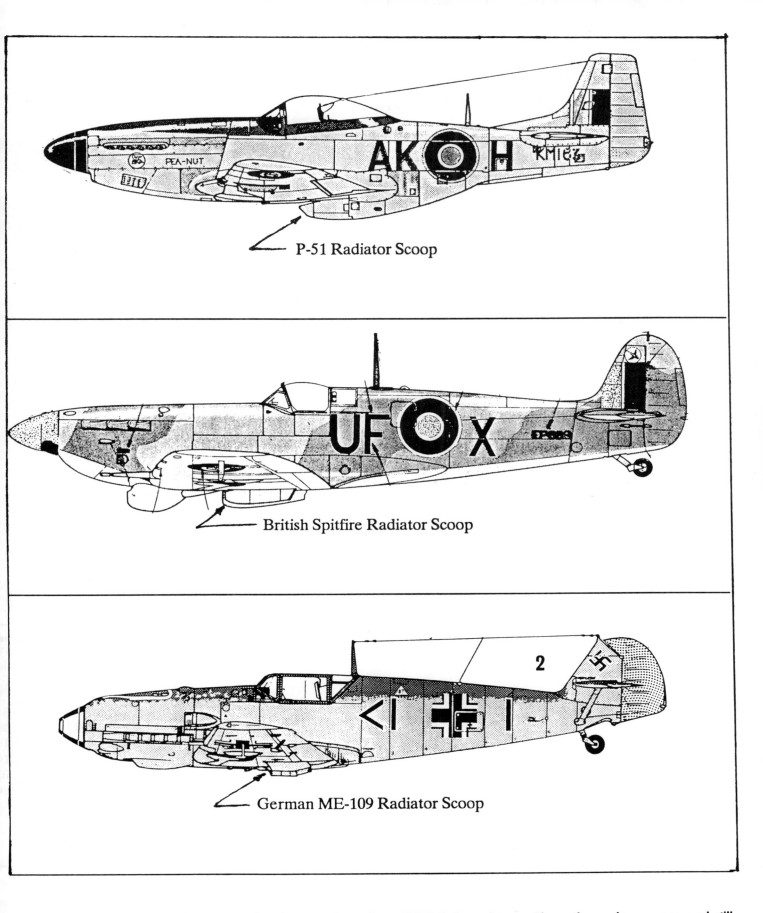

P-51 Radiator Scoop

British Spitfire Radiator Scoop

German ME-109 Radiator Scoop

Check the three different radiator scoop locations on these three WWII fighter planes. These three planes were and still remain very, very successful, and therefore merit consideration when designing a water-cooled radiator location for an experimental airplane powered by an auto engine. DWGs courtesy U.S. ARMY AIR CORPS, WW-II I.D.

A typical Kitfox airplane with a water cooled Rotax engine makes use of this belly-mounted radiator in a scoop for proper cooling. Hanging the radiator out in the air without the scoop would not cool the engine.

RADIATOR INSTALLATIONS THAT WORK

The number of airplanes that are successfully flying with water-cooled auto engines for power is very substantial. Therefore, it makes good sense to copy radiator designs that are proven. This chapter will provide several choices in radiator installation designs.

PREFERRED RADIATOR LOCATIONS

For water-cooled airplanes that go fast, you can't beat the P-51 Mustang World War II fighter plane. The radiator on this fantastic airplane is mounted under the belly of the airplane, in a relatively small scoop, and wind tunnel and flight tests have proven that the hot air coming out of the duct behind the radiator actually adds a tiny bit of thrust to the airplane.

Other proven designs are the British Spitfire World War II fighter plane that uses two smaller radiators, mounted in scoops under the wing near the landing gear. The "Prowler" auto engine powered airplane built by George Morse (and pictured several places in this book) also uses two small, under-wing mounted radiators.

Steve Wittman, designer of the Buick V-8 powered Tailwind airplane, mounted a full-size car radiator above the inverted V-8 engine, in a flat position. Air flows into the radiator through the normal engine cowl openings, and exits upward through louvers that contribute to rearward air flow of the exhaust air. More recently, Rick Schneider of Eagle River, Alaska, installed a Mazda 13B rotary engine in his Bushmaster, cub-like airplane, and mounted the radiator flat, on top of the engine, and let the exhaust air exit the engine compartment via louvers in the top cowling, much the same as Steve Wittman did with his Tailwind radiator installation.

Fred Geshwender mounted a full-size car radiator flat, under the Ford V-6 engine in one of his Ag plane conversions. As in the Schneider

Jess Meyers, of Las Vegas, Nevada, eventually mounted this radiator and P-51 style scoop on the belly of his Buick V-8 powered Globe Swift airplane. Note the surface mounted aluminum pipe that carries anti-freeze from the engine to the radiator and back.

and Wittman radiator installations, the radiator exhaust air exited via louvers, but in the bottom of the nose cowling. In the 1935 Funk-certified airplane, the manufacturer mounted a ar radiator vertically, on the pilot's side of the Ford 4-cylinder engine, and let the hot exhaust air from the radiator exit the cowling through louvers in the side of the cowling.

All of the above-mentioned radiator installations work well and cool the engines adequately. Because the radiator can be placed almost anywhere, and engine coolant plumbed to and from them, there are numerous other radiator installations that are feasible.

This highly efficient belly mounted radiator is on a Chevrolet V-6 powered Velocity RG airplane built by Timothy England, of North York, Ontario, Canada. Note the very narrow inlet of the radiator plenum, probably 2" high by 18" wide. Tim says that, initially, the Chevy engine ran TOO COOL until he installed a 190°F. thermostat in the engine.

RADIATOR DON'Ts

- Don't try to turn the cooling air more than 90° to the direction the airplane flies.
- Don't place the radiator where engine exhaust manifold heat can enter the inlet air flow.
- Don't use as much radiator capacity as your car needs.
- Don't mismatch the auto engine water inlets and outlets to the radiator inlets and outlets by more than 50%.
- Don't put the radiator fill cap lower than the highest point in the cooling system.

RADIATOR DO'S

- Do design the radiator ductwork for as smooth and straight an air flow as possible.
- Do make the radiator inlet scoop at least 25% smaller to 50% smaller than the air outlet, so the hot, expanded air can exit easily, and so the inlet air will not stagnate.

- Do shield engine heat and exhaust system heat away from the incoming radiator air.
- Do incorporate a radiator filler cap in the highest point in the water system, such as in a remote mounted header tank, so no air will be trapped in the cooling system.
- Do incorporate a plastic overflow tank like your car engine uses. It is easy to do visual water/coolant level checks by just looking at the green antifreeze in the overflow tank.

WATER COOLING SYSTEMS

Designing a suitable cooling system for your water-cooled auto engine can be a lot of fun, but doing it properly can make or break your auto engine conversion. If you don't get good water circulation and enough air flow through the radiator or radiators (plural), the engine will overheat and possibly cause a forced landing.

Designing a suitable water cooling system can be very easy to do as long as you observe a few basic rules of physics.

- Impeller type water pumps will not pump air, so keep the water pump low enough that it is always submerged in water.
- Don't build in air traps in the water system. Install airbleed pet crocks or valves to bleed air out of high points if there are any high points not open to air (such as filler caps).
- Don't try to over-complicate the cooling system. You should be able to adapt existing auto or motorcycle radiators for less than $60.00!
- Don't try to make the natural airflow of the airplane flying forward at 100 MPH to 200 MPH, do crazy things like flow from the firewall forward or do 180° or 360° turns. Air exiting the radiator can go up, down or even out one or both sides of the airplane, but not in at the middle of the airplane and out at the front!

75

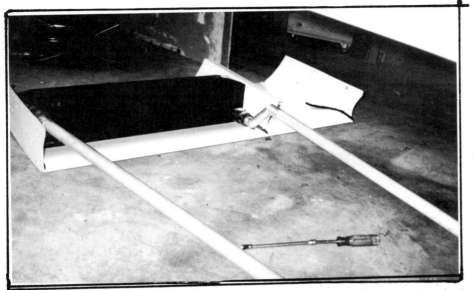

Many radiators can simply be hung under the fuselage in a removable plenum chamber/scoop like this one for my Grumman Traveler 4-place airplane with its Buick V-6 engine conversion. I used 1 1/4" diameter 6061-T6 aluminum tubes to carry the antifreeze. The radiator was custom made from brass.

P-51 RADIATORS

If you simply can not find a suitable place for a radiator in your engine compartment or in the wing, a P-51 scoop will always get the job done. The important thing is to design a plenum chamber ahead of the radiator. This plenum chamber should be at least 12" deep, 6" tall, and 24" wide for most V-6 and V-8 engines. The inlet should be only 50% as tall as the outlet of the scoop. This is because hot air off the radiator will expand as it leaves the radiator.

Two excellent examples of P-51 type radiator scoops are Jeff Ackland's Legend on this page, and

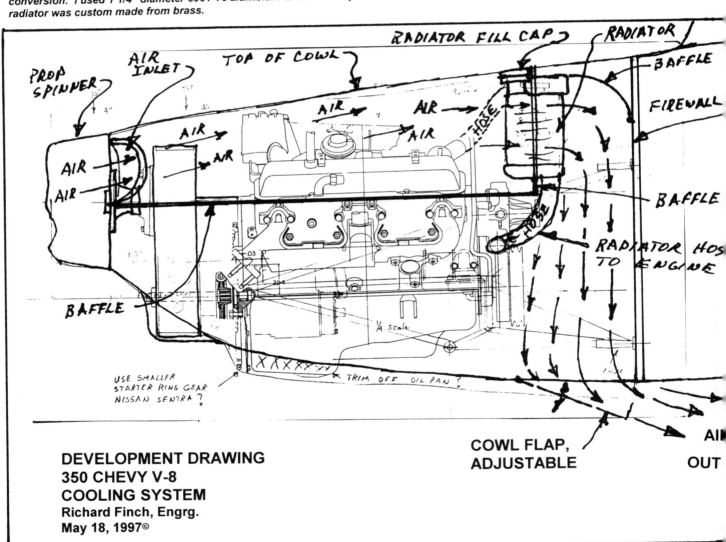

DEVELOPMENT DRAWING
350 CHEVY V-8
COOLING SYSTEM
Richard Finch, Engrg.
May 18, 1997©

The rough sketch above shows that a suitable radiator can indeed be mounted in the engine compartment of a Jeffaire Baracuda airplane with a 350 horsepower Chevrolet V-8 and a plans-built Finchbird belt drive. There is even room in this engine compartment for a Freon air conditioning system, just like they use in street rods. The builder of this conversion is Brad Sunde, of Fergus Falls, MN.

Timothy England's Velocity on page 75. Design your radiator like either of these two radiator scoops and you will not go wrong. Remember, copy the winner. Remember that water seeks its lowest level; air rises.

YOU MUST INCORPORATE A SURGE OR HEADER TANK

The only sure way to guarantee that your auto engine water pump does not cavitate or suck air is to mount a tank or reservoir at the highest point in the cooling system. This will assure that if no leaks occur and the radiator does not boil out the coolant, you will always have water flow.

HEATING THE COCKPIT

With a water-cooled auto engine powering your airplane, you have the ability to provide for plenty of heat. You can buy a used heater radiator from an auto salvage yard and mount it anywhere there is sufficient room For a picture of several bare auto heater radiators, look at the picture below.

You can either let outside air blow through the heater radiator core, by means of a duct hose, or add a small electric fan to simply recirculate the air through the heater and back out into the cockpit. Make a simple box to enclose the heater radiator so you can connect a 1½" to 2" air hose to it for windshield defogging. Of course, a valve in the hose is necessary to stop the hot air when you don't need it for defogging. You can make a suitable heater core box from .020" aluminum sheet, fiberglass, or Royalite plastic sheet. Seal the area where box touches the heater radiator with R.T.V. or other aircraft sealant so vibration won't wear holes in the radiator.

Jeff Ackland's Big Block Donovan Chevy powered Legend airplane makes full use of the P-51 Mustang style radiator scoop. Here you see the exhaust end of the radiator. The radiator is custom-made by Griffin Radiator.

Shirl Dickey uses a NACA duct scoop on the belly of his E-Racer canard airplane powered by this Chevrolet V-6 engine. In this picture you can see the large oil cooler radiator laying at about 25°, just below the radiator expansion tank. Shirl is in the background.

I would put a manually operated water shut off valve in the water inlet hose so the hot engine water can be shut off in warm weather. Use rubber or nylon grommets where the rubber heater hoses run through the fire wall. You can buy vacuum operated water shut-off heater valves that would be very convenient. A simple open-close switch on the instrument panel would turn the heater water "off" or "on." While you are at it, investigate the various vacuum diaphragms the automobiles employ to open and close air vents to the heater and defroster air.

PHYSICS...AGAIN!

When the OX-5 V-8 engine was designed in about 1920, we did not have pressure radiator systems in our cars, and therefore we did not have pressure radiator systems in our airplanes.

The system in those days was to fill the radiator (on your car or airplane) with water and then just expect it to slowly evaporate through the open radiator cap as you drove or flew. At every fuel stop, you had to add water to the radiator because of evaporation.

In order to keep the water flowing, even when the radiator was low on water, the solution was to place the water pumps as low as possible in the cooling system. Certain engines such as the 1925 Ford Model T did not even have water pumps, but depended on hot water rising to the top of the engine and radiator, and on cooler water falling to the lower engine inlet.

The water pump on the OX-5 V-8 engine is mounted as low as possible to avoid air traps and overheating when the non-pressure radiator ran low on water. This is a good feature, even for pressure radiator systems.

The same radiator physics applies to modern pressure radiator systems. If the water pump is located at the top 90% of the coolant system, it will not pump water if the water level falls below 90%. Therefore, keep the water reservoir as high in the system as possible.

Ed Lubitz, of Ontario, Canada, said he could do a complete auto engine conversion in one week-end, and this photo shows why. The radiator has a simple sheet metal plenum chamber that he could fabricate in an hour or two. And, the flex hoses solved his installation problems. The engines were Ford Escort 4 cylinder units, not modified at all.

MORE PHYSICS

Some very vocal, early experimenters with auto engines in airplanes, believed, and shouted to anyone who was interested, that airplanes with auto engines needed bigger radiators than cars with the same engines.

But, the SUCCESSFUL experimenters have all come to the same conclusion about radiator size. If you have proper air flow through the radiator, and no air bubbles in the system, the airplane can use 1/2 to 1/3 the size radiator that the same engine in a car would require.

WHY?

Because the airplane flying at 100 to 200 mph, is passing 3 or more times as many cubic inches of air through the HEAT EXCHANGER than the car is at 65 mph. Again, copy the winner, not the also-rans.

Flywheels, Flexplates

CHAPTER

10

◆◆◆◆◆◆◆◆◆◆◆◆◆◆◆◆◆◆◆◆◆◆◆◆◆◆◆◆◆◆◆◆◆◆◆

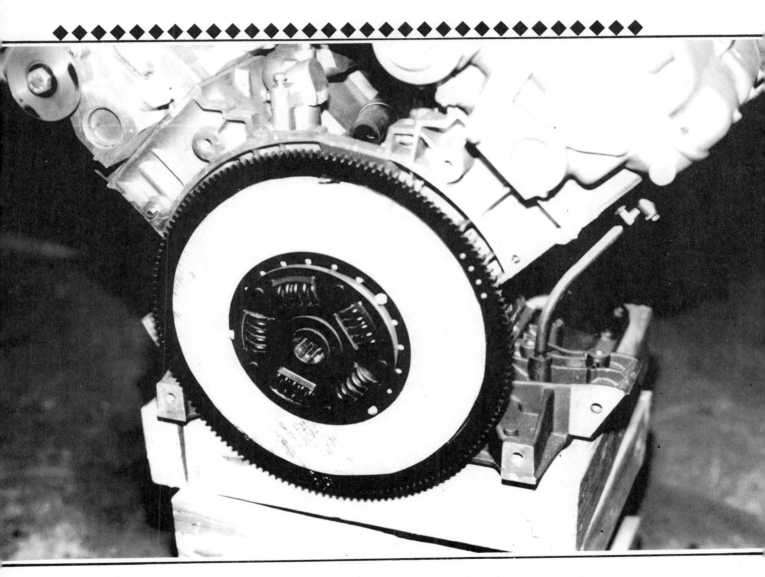

In this picture, you can see a plywood mock up for a FINCHBIRD V-6 flywheel spring hub, flex plate adapter. When it is completed from steel, it will weigh 10 lbs. The engine is an all-aluminum Renault V-6.

FLYWHEEL OR NOT?

Most auto engines will run without a flywheel of any kind. My Buick V-6 engine in my Grumman 4 place airplane will start with only a flex plate, run up to 5000 rpm and then settle down to a very smooth idle at 600 rpm. In fact, the 220 horsepower, 231 cubic inch Buick V-6 will run so smooth without a flywheel that you can set a full glass of water on top of the aircleaner and it will not be spilled from idle speed all the way up to 5,000 rpm and then back to idle again.

Of course, you need a starter ring gear or an automatic transmission flex plate in order to start the engine, but the flex plate only weighs 4 to 5 pounds at the most and does not contribute to flywheel effect. The auto engine crankshaft provides the inertia to

79

The Precision Arial Services 164 cubic inch Corvair 6 cylinder engine is set up for direct drive, and uses a smaller than usual flywheel/flexplate from a Nissan Sentra. The small diameter flywheel makes the engine frontal area smaller for better cowling streamlining.

This Triumph V-8 (Buick derived) engine uses a clutch hub to drive the shaft and lower sprocket of the plans-built FINCHBIRD PSRU. Neoprene or rubber bushings could be used to dampen the torsional vibration pulses to a gearbox, chain drive, or belt drive.

David Stangland Photo

keep the engine running and running smoothly. But it is not perfectly smooth.

DESTRUCTIVE VIBRATIONS

Each time a piston comes up on the compression stroke, and goes down on the firing (power) stroke, small but significant vibrations shake the entire engine and whatever frame it is mounted in. Balancing the engine helps tremendously, but the actual firing stroke of any internal combustion engine causes vibrations. These vibrations can shake gear and chain drive prop speed reduction units (PSRU) to destruction in time.

Many inventors of cog belt drive reduction units expect that the vibrations will be absorbed in the rubber teeth of the drive belt, and this is true to a great extent. It is true that the rubber and fabric drive belt does soften the vibrations between the engine and the prop to a major extent. But there have been a significant number of cracked flex plates and even broken crankshafts on engines that run flex plate and no flywheel. A flywheel is good to have for the purpose of helping to smooth out the vibration caused by the firing stroke, even with a cog belt drive.

HOW MUCH FLYWHEEL?

Fred Geschwender recommends a full-weight flywheel to absorb the firing pulses on his

Morse chain drive engines. George Morse runs a significant weight flywheel on his aluminum V-8 engines that feature front mounted gear box prop reduction units. A stock Buick V-6 car with a stick shift transmission uses a 40 pound flywheel to assist in smooth take off from stop signs. The two airplane engine conversions mentioned in this paragraph feature 10 to 15 pound flywheels. So, which is better?

The 40 pound flywheel is obviously too heavy for airplane use, and it is way too heavy for even street rod use in a car. Six to eight pounds of flywheel weight in the right place will help smooth out a lot of un-wasted vibrations in your auto engine conversion. The right place for flywheel weight is at the outer circumference of the available space between the engine and the prop drive unit, on the same plane as the flex plate.

IDEAL FLYWHEEL

The most efficient flywheel is the one that is thin in cross section near the center or hub area of the flywheel and thicker near the outer circumference. the illustration in this chapter shows an efficient and effective flywheel design that will work well on most auto engines.

CUSHION HUBS

A rather simple and effective way to cushion the vibrations between the engine and the prop drive unit is to incorporate an automotive or boat cushion drive hub in the flywheel unit. Most automobile stick shift cars use a

Look closely and you will see the counterweight between the engine and the flywheel/flexplate on this Mazda Rotary engine. All Mazda Rotary engines must have this removable counterweight to prevent severe vibrations when the engine is running.

multiple-spring cushion hub in the clutch disc to soften the clutch engagement and disengagement and to take out the engine firing pulses from the transmission and driveline.

Boat inboard-outboard drive units incorporate a similar cushion hub between the engine and the propeller gearbox, except that rubber is used rather than multiple springs as you find in stick shift car clutch hubs. A rubber cushion hub will soften vibrations better than springs, but the spring hubs are much easier to find and to adapt, and they cost $25 to $50, a very reasonable price.

Check the pictures in this chapter for ideas on how to adapt a light and effective flywheel to a cushion drive hub.

PATENT A PROP DRIVE

For several years, I had experimented with the idea of a prop drive unit that will fit any brand auto engine by simply using an adapter plate. In this design, one prop drive unit could be manufactured in large quantities, making for a less expensive and a more proven and dependable design. This drive unit could be designed and built in three different horsepower capacities, one from 100 to 200 horsepower, one from 200 to 300 horsepower, and one from 300 to 500 horsepower. The input drive shaft would simply fit into the cushion hub spline at the flywheel end of the auto engine, and a space adapter could allow one prop drive unit to be fitted to Ford, Chevrolet, Buick or other brand of engine in the specific horsepower range.

But I didn't apply for a patent on this idea, and since it is now published in this book, it becomes public domain and that means that anybody can build these "universal gearboxes" and sell them; so do it. If race car people can build a two

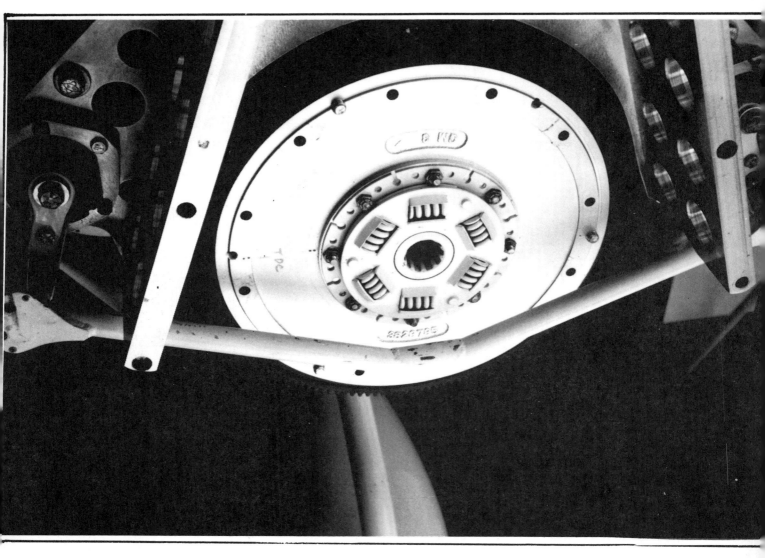

With the gearbox PSRU removed from my Buick V-6 engine, you can easily see how a spring cushion hub has been adapted to the steel flywheel and then to the Buick 160 tooth starter ring gear and flex plate. The entire assembly was balanced prior to installing it on the engine. Balancing this assembly cost $25; well worth it!

speed racing transmission and sell it for $895, then it is possible to build airplane engine prop gear boxes and sell them for $895 also. The sketch in chapter 13 will illustrate the idea of the universal fit prop reduction drive unit that would be easy to design and sell.

HEAVY FLYWHEELS

My Dad's 2 cylinder John Deere farm tractor had a very heavy flywheel to smooth out the power pulses delivered by the 360° firing order 2 cylinder engine. The flywheel weighed over 100 pounds. If you ever heard a 2 cylinder John Deer tractor turning around at the end of the field, you could actually count the number of firing strokes it made as it went "pop, pop, pop". We used to call those tractors "Popping Johnny's" because of the widely spaced firing strokes.

If you intend to completely smooth out the firing strokes from a piston engine, you would need a flywheel that is almost as heavy as the entire engine! The FAA tells me that a 4-cylinder certified Lycoming engine is the most vibration prone engine in General Aviation. Recent solutions to the certified engine vibration problem has been to add a very large and heavy vibration damper to the crankshaft, behind the propeller.

HAPPY MEDIUM

You need some flywheel effect, more than a 5 pound flex plate provides, but surely much less than a 100 lb. + tractor flywheel. If you incorporate 10 to 15 pounds of flywheel, you will significantly reduce the firing stroke vibrations. The flywheel, flex plate, clutch hub assembly shown on the Buick V-6 engine on this page weighs 16 pounds. And it works well.

Cog Belt Drive Designs

◆◆◆◆◆◆◆◆◆◆◆◆◆◆◆◆◆◆◆◆◆◆◆◆◆◆◆◆◆◆◆

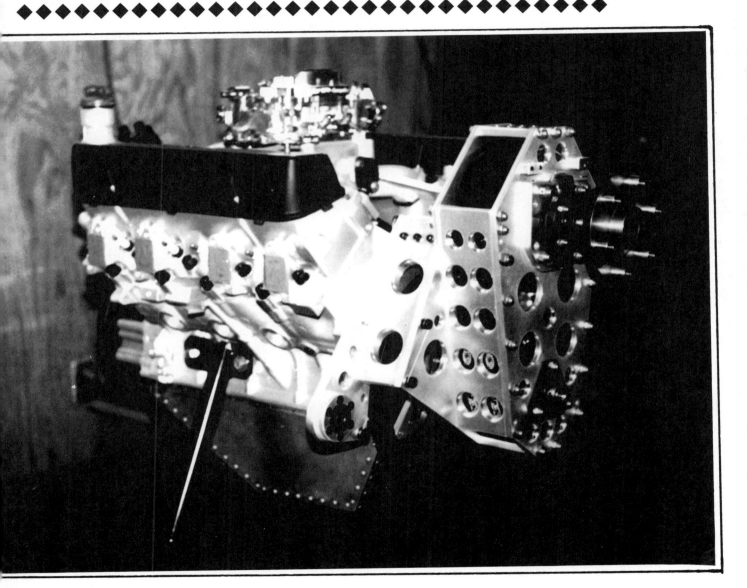

This is a plans-built belt drive PSRU mounted on a low-time, all aluminum V-8 engine. The builder, David Stangland, of Albion, Ohio, did an excellent job, making the cog belt drive unit show quality.
The belt is a Gates Poly-Chain unit, rated for 315 horsepower. The engine will likely produce 175 to 200 horsepower, so the PSRU is very adequate. PSRU weight is 62 pounds.
David Stangland Photo

CH. 11
COG BELT DRIVE DESIGNS

PLANS BUILT BELT DRIVES

Shown on page 83 and 84 is a plans built belt drive for a Triumph V-8/Buick V-8 engine. Similar belt drive plans are also available for Chevrolet V-6, V-8, Buick V-6, Volvo V-6. Buick 3800, and Mazda rotary engines. For plans, contact **FINCH ENGINEERING**. See page 152.

EXAMPLE OF COG BELT WEIGHTS & COSTS

David Stangland, Albion, Indiana, built the PSRU pictured on page 83. The following figures were submitted by the builder:

COSTS:

- Reduction Unit Dwgs......................$102.00
- 6061 T6 Alum. Plate......................$264.00
- 4140 Steel Shaft, Splined...............$ 95.31
- 28 Tooth Sprocket.......................$107.38
- 43 Tooth Sprocket.......................$137.21
- 14 mm Cog Belt..........................$178.97
- Dana Axle, Prop Shaft...................$ 60.00
- Crankshaft Spacer.......................$ 15.00
- Splined GM Hub..........................$ 15.75
- Input Shaft Lock Bushing................$ 31.93
- Prop Shaft Lock Bushing.................$ 33.14
- Bearings, 4.............................$104.40
- Hardware................................$138.26

Total: **$1,283.95**

WEIGHTS:

- Prop Shaft..............................8.67 lbs.
- Input Shaft.............................1.58 "
- 28 T Sprocket, Machined.................6.98 "
- 43 T Sprocket, Machined.................8.40 "
- Aluminum Housing Plates.................19.57 "
- Bearings, bolts, hdwe...................17.01 "

Total Weight: 62.21 lbs.

- Fly Wheel Assembly:.....................11.71 lbs.

This off-the-shelf Gates Poly Chain Cog Belt Drive unit is very heavy when it comes in the box, but careful lathe work will remove more than half the weight of the steel sprockets.

SOURCES OF PARTS

The cog belt, sprockets, sprocket hubs and hardware listed on this page, came from a local business that was listed in the telephone book under "Bearing Supplies". The axle shaft came from a new car parts department, and the aluminum plate came from a local business listed in the Yellow Pages under "Metals, Aluminum, Brass & Copper".

PLANS BUILT

You can buy the parts and materials and build your own Cog Belt Drive PSRU from plans that you can buy from several sources; the cost will be about 1/2 to 1/3 the cost of a ready made unit. The photo on page 83 shows an excellent example of a plans-built Cog Belt Drive unit mounted on a nearly new British Triumph 215 V-8 engine.

This plans-built belt drive unit weighs about the same as several ready-made drive units. For further information, see Chapter 19, "FINCHBIRD PLANS". The plans sell for $100 per set. Buick V-6, Chevy V-6, V-8, and Renault V-6 engines can be converted to belt drive with these plans.

The engine pictured here is a 2.7 liter Renault V-6 with a plywood mockup of a belt drive PSRU. For all first-time, "one-of" designs, plywood should be used to make the mock-up.

MOST AVAILABLE PSRU

Because of the relative ease of obtaining the correct parts, and the relative ease of manufacturing a cog belt PSRU, there are more of these units available to the auto engine converter than all other units combined, chain drive, gear drive, and direct drive.

Anyone who can build an experimental airplane from any material, fiberglass, aluminum, steel, or wood, can build a cog belt drive PSRU unit. Of course, you must have access to a lathe, milling machine, band saw, drill press and metal working shop tools.

For the builder who works only in wood or composites, he can usually sign up for an evening class in machine shop at his local community college or even high school. And then there are builders who work in industrial businesses where these metal working tools are available for use by employees.

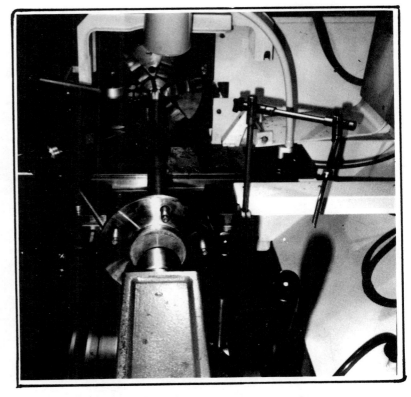

Here you are looking at a Chevrolet Caprice rear axle shaft, 1975 to 1994, chucked up in a lathe for turning and making into a belt drive prop shaft. The tensile strength of this axle shaft is 190,000 psi, and the diameter is 1.5", so the shaft is very strong.
Turning the shaft is easy to do at 500 rpm, using a disposable carbide tip cutter that costs less than $10. Any high school metal shop can turn down this type of axle shaft.

HOW TO SELECT A COG BELT DRIVE

Look at the many cog belt drive systems shown in this book. Some are marginal for the horsepower they are expected to transmit to the prop, and some are extremely overbuilt.

In racing, if we wanted to finish 1st or 2nd place, we copied what the *winner* was doing (building) if we didn't have the time and money to do our own experimenting. Race car drivers of *all* race classes still do that. When the winner puts a turbocharger on his race car, everybody else (who is smart) does the same thing. It sure saves a ton of money to let someone else do the experimenting.

Some of these cog belt reduction drive systems are flying constantly with no problem, and others are just too complicated to be affordable by the average homebuilder.

If you can build an airplane, you can probably build your own cog belt drive. You can get help from your locaL industrial supply company or your local bearing

One of the very common uses for Gates Poly chain cog belt drive is shown here in this rear drive on a 68 horsepower Harley Davidson motorcycle. It runs in the dirt, water, mud and gravel, and lasts forever. It will also last in your airplane.

supply house. You can also write for a cog belt installers' handbook. Ask for"

GATES POLY CHAIN
G.T. DRIVES CATALOG
Gates Rubber Company
990 South Broadway
P.O. Box 5887
Denver, Colorado 80217

One of the nicest belt drive conversions ever is this very clean unit mounted on a Honda 4 cylinder engine. This unit is made by Avtec of Montana. Expected horsepower is about 125 hp.

DESIGN IT YOURSELF

Using the photos in this chapter, you can design your own PSRU belt drive unit, to fit engines not normally converted, such as V-4 and V-6 outboard boat engines. I have completed and flown an Evinrude V-4 powered Scorpion Two helicopter with a belt and chain drive, and it is a very viable thing to do. Some outboards produce over 200 horsepower at 5,500 rpm.

HOW TO CHOOSE A PROPER REDUCTION RATIO

1. Determine the desired prop rpm, say 2,650.
2. Determine desired cruise rpm of the engine. If the red line is 9,500, cruise is probably 66% of that. Divide 9,500 by 3 for 1/3, then multiply that answer by 2, for 6,333 rpm, (or use the equation: 9,500 x .666=6,327 rpm).

3. Now divide the desired engine cruise rpm (6,327 rpm) by the desired prop rpm at cruise (2,650 rpm), and the answer is 2,387.

4. This means you want a reduction ratio of about 2.387 for your belt or chain drive.

5. Say you are going to use a motorcycle #40 chain and sprockets, and the big sprocket on the prop shaft has 40 teeth. Divide 40 teeth by 2,387, and the answer is 16.7574, or round it off to 17 teeth.

6. So the engine sprocket would be 17 teeth, and the prop shaft sprocket would be 40 teeth. If you divide 40 teeth by 17 teeth, your answer would be 2.3529411 for a rounded off ratio of 2.4 to 1.

7. If your prop should turn at 2,650 rpm for cruise, multiply 2,650 rpm by 2.3529411, and the answer is 6,235 rpm. This is the rpm the engine would be turning at cruise speed.

NOTE: The above rpm figures are for a rotary engine. Most piston engines will have a maximum power output at 5,200 rpm, not 9,500 rpm.

SUITABLE RATIOS

After 25 or more years of experimenters flying belt drives, the ratios that work are pretty well established. There are some exceptions to the standard, such as 7 foot, 3 blade props for STOL airplanes that need a higher numerical ratio, and for older engines that will not tolerate continuous high rpm operation, that need a lower numerical ratio.

1.63 to 1.75 to 1 RATIOS

This numerical drive ratio range works very well for sport-type airplanes that fly at 125 to 200 miles per hour. Three blade props work better with these ratios than 2 blade or 4 blade props do. These ratios provide for the maximum horsepower at take-off (5,250 rpm divided by 1.75=3,000 prop rpm) and good power at cruise (3,850 rpm divided by 1.75 = 2,200 prop rpm).

One of the two best cog belt drive units available is this one on Elwyn Johnson's 4.3 liter Chevrolet V-6. You can build this unit, ready made, for $3,000, and bolt it to your Chevrolet V-6, V-8, or Ford V-6 engine.

Another example of the excellent Northwest Aero Products cog belt drive PSRU; this time it is mounted on Tim England's Velocity RG in a pusher configuration. You also get a good look at Tim's electric in-flight adjustable IVO Prop in this picture.

Corvair engines are excellent choices for low budget airplane projects. You can still find "rebuildable" used Corvair engines in almost every town in the USA, and in many countries in Europe. All parts needed to completely overhaul any Corvair engine are available from the very large CLARK'S CORVAIR PARTS catalog. If you plan to rebuild and fly a Corvair, you will also want a copy of the 8th edition of "HOW TO KEEP YOUR CORVAIR ALIVE", by Richard Finch, available from Motorbooks, International, and from CLARK'S CORVAIR PARTS.

This picture really looks like a LaFayette Escadrille photo from 1916 - 1918, but it was actually taken in 1994. Rich Dietrich, of Morrice, Michigan, built this N-12 replica and installed a 2 to 1 Gates Poly Chain reduction drive that he designed and built himself. Rich and his wife have dressed in WW I style clothes to take this picture, and it sure LOOKS like WW I.

Clark's Corvair Parts, Inc.
Corvair Engine Parts
Route 2, 400 Mohawk Trail
Shelburne Falls, MA 01370
1-413-625-8498 (fax)
1-413-625-9092 (phone)

When you calculate these 1.63 to 1.75 PSRU ratios, you will see that they let your engine operate at its most efficient rpm range.

Because the teeth and the belt engage each other more often at even ratios, 1.0 to 1, 2.0 to 1, 3.0 to 1, & etc.; avoid these ratios to avoid accelerated wear patterns.

HOME MADE BELT DRIVES

On this page, you see two pictures of an excellent home made belt drive system that works exceedingly well for 4 and 6 cylinder opposed engines such as the VW and the Corvair. The Corvair is the only American made engine that turns "backwards" compared to aircraft prop rotation.

When you use a long prop shaft mounted on top of the engine like this conversion, you have a choice of driving the prop from either end of the shaft, thereby making it possible to use clockwise rotation props. (All dimensions and rotations in all airplanes are expressed as seen by the pilot when he is sitting in his pilot's seat.)

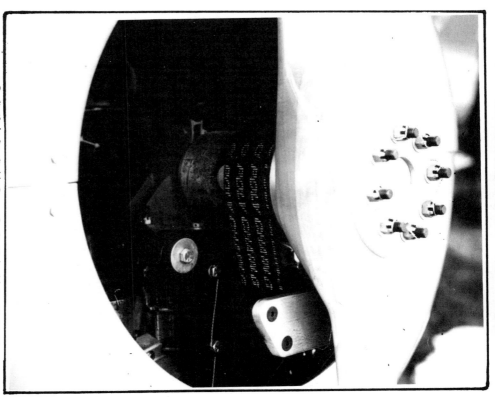

In this picture, you can barely see the 8 MM Gates Poly Chain GT belt, operated by a 28 tooth sprocket on the engine, and turning a 56 tooth sprocket on the 3 inch diameter prop shaft. The prop shaft is a 1/8 inch wall tube that runs the length of the 164 cubic inch Corvair engine, mounted on two bearings; 5.82 inches above the crankshaft center lines.

This fuel injected Subaru Legacy engine has been fitted with dual cog belts, 8 MM tooth spacing and a strong, but heavy, frame of aluminum. Formula Power Company; of San Jose, California designed and built this cog belt drive.

The belt drive unit in this picture is a modified version of the Formula Power unit in the previous picture. Note the lighter aluminum housing that has been milled on a CNC milling machine. The new unit is lighter and stronger than the original unit.

One of the more successful auto engine converters is CAM of Canada. This CAM 100 engine, firewall-forward kit is based on a 4 cylinder Honda engine, is very compact and dependable, and features 4 separate ignition coils mounted on the firewall. Horsepower outputs of 100, 125, 150, and 175 are available from CAM.

8 MM BELTS

The designation 8 MM applies to the spacing between the teeth on the sprockets and the belt. These belts come in several widths, from the 1 1/4" sizes shown on this page to the much wider 8 MM belts used in the drive units shown on page 87. Those two belts are about 5" wide.

When you use the Gates Catalog to pick the size of the belt for the horsepower you expect to get from your engine, you will be shocked when you see that a 8 MM belt, 1 1/4" wide, (30 MM) is listed at a mere 28 horsepower! But don't let the catalog figures discourage you. The 5 + " wide 8 MM belt used on Chevy V-6 & V-8 engines is only rated for 80 horsepower. This means that the Gates factory engineers are ultra-conservative in their horsepower estimates.

DEPENDABILITY

Even though the factory catalogs and specification guides show a relatively low horsepower capability for 8 MM belts, many inventors have obtained excellent dependability from this size belt. One notable example is the Chevrolet V-6 and V-8 belt drive conversions sold by Northwest Aero Products.

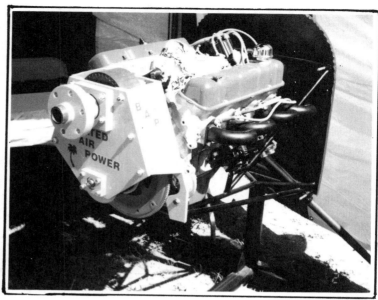

This display engine and belt drive PSRU was on display at the 1991 Copperstate Fly-In at Prescott, Arizona. It is the Belted Air Power Buick V-8 engine conversion that Jess Meyers has flown successfully since 1983.

This belt drive PSRU is made by Reiner Hoffman, of Stratus, Inc., in Washington State. It is mounted on a Subaru EA-81 engine and flies in an Avid Flyer 2 place airplane. This conversion is very dependable and has flown for many years.

FLEX PLATES ONLY

For over 25 years, auto engine converters have relied on a 4 to 5 pound flexplate to provide a starter ring gear, and no flywheel at all. For the most part, this arrangement works OK.

But you may experience propeller failures if you use anything but a one-piece prop. There have been a number of prop failures from laminated wood props coming apart, due to the firing pulses from the engine. One solution is to incorporate a suitable flywheel to smooth out the power pulses.

COG BELT "TBO"

Years of experience indicate that a cog belt flown in an airplane will last about the same number of months that a tire will last. If you replace your tires every 400 to 500 hours, you should also replace the belts at the same time.

Heat and direct sunlight affect the life of cog belts. so protect the belts from those elements. A typical 8 MM cog belt sells for $75 to $100.

Here you see a Porsche air cooled engine with dual 2 barrel Weber Carburetors, operating a tubular prop shaft via an 8 MM belt drive at the rear of the engine. This arrangement turns the prop counterclockwise. If the engine were turned around, like in the Rich Dietrich Newport-12 Corvair engine on page 88, the prop would turn in the standard, clockwise direction.

Chain Drive Designs

CHAPTER

12

This prototype Morse Chain Drive unit mounted on a bare 4.3 Chevrolet V-6 engine block, is the way to go for trouble free PSRU design. Milo Burroughs, Aero Kinetics, Inc., of Yelm, Washington, makes this kit that you assemble yourself from parts that he furnishes plus off-the-shelf parts that you buy from your local industrial bearing supply.

FUTURE PSRU'S

The Morse Type, silent chain drive seems to be the PSRU of the future. Typically, in road vehicles, the chain drive will last as long as the vehicle. They do wear out, but a 100,000 miles or more. A major advantage to silent chain drives, is the ability to perform at high rpm, as much as 8,000 rpm.

In the prototype chain drive unit shown on this page, several necessary components are missing. There is an aluminum housing that fits between the two aluminum plates to contain the lubrication oil. There is also an oil spray bar that sprays oil on the chain, and there are oil lines to carry the oil to and from the engine.

Also needed in this chain drive unit are suitable rubber engine mount vibration isolators. And the prop flange will also require prop bolt drive lugs to engage the prop. This photo is of a display engine and it is deleting parts to more clearly show the chain.

91

ADVANTAGES OF CHAIN

This chapter is about the multi-link silent chain drives, in case you were thinking about bicycle chain or even log chain! But if you have the opportunity to see the Wright Brothers Kitty Hawk power airplane, be sure to notice that is has two propellers turned by one engine, through motorcycle size roller chain. That is, the chains are the size we are familiar with that drive motorcycle rear wheels. Chains are often used because they are inexpensive and convenient, and because they work.

PROP SPEED REDUCTION CHAIN DRIVE UNITS

Here are some reasons to consider a silent chain transfer case rather than a belt drive unit or a gear box.

The first advantage for the homebuilder or the prototype transfer case builder, is that the input and output shaft center distances are less critical in a chain unit. In a gear unit, too much backlash can cause excessive gear wear, and too little backlash can cause the gears to overheat and gall. A gear box must be machined to exact dimensions whereas a chain drive case can tolerate several thousands of an inch mismatch.

Other advantages to silent chain drive are:

■ Up to 99% efficient power transmission

■ Fewer parts are required in a chain drive

■ Chain drive cases are lighter than gear drive cases

■ Chain bearing loads are in compression. Gear bearing loads are out- ward, causing more case stress

■ Chain drives are better for high speed operations from 3,000 rpm up to 7,000 rpm, and even higher if necessary

Fred Geschwender of Lincoln, Nebraska, makes this Morse Hy-Vo chain drive unit for high horsepower engines like this 620 cubic inch V-8 in the Legend airplane.

This Fred Geschwender chain drive PSRU in the Legend airplane uses 2" wide x 1/2" pitch chain, driven by a 2 to 1 ratio and operates a hydraulic prop. Note the exceptional housing casting quality, developed over a period of 25 years by Fred Geschwender.

This picture of the Geschwender chain drive shows the line for oil pressure in, and the big 90 degree fitting on the bottom of the housing where engine oil is scavenged out of the unit. Fred has always incorporated spray bar lubrication and provisions for constant speed props in his chain drive PSRU's.

Ray Ward, of Houston, Texas, has flown his BD-4 airplane with a small block Chevrolet V-8 engine for over 10 years. In 1996, he switched from a Universal Engineering chain drive to a Geschwender chain drive. Ray has been known to outrun a turboprop Beech King Air! That means that this airplane can go faster than 250 mph.

■ A chain drive is "elastic" and therefore tends to cushion the vibration loads

■ Millions of units and millions to billions of miles of dependable service in transverse drive auto engines and in 4 wheel drive transfer cases make for a tremendous dependability study

HORSEPOWER CAPACITIES

When you consult a manufacturers' catalog to get the facts for designing a chain drive prop speed reduction (PSRU) case, you will be surprised and shocked to see how low the factory horsepower and rpm figures are. The same thing is true if you are designing a belt drive reduction case. The manufacturers' specifications are VERY, VERY conservative. I will give you four examples:

MOTORCYCLE CHAIN

A number 50 motorcycle chain can handle 150 horsepower and speeds to 100 to 125 miles per hour. In racing applications, we see over 200 horsepower out of 900 cc motorcycle engines and speeds of over 170 miles per hour.

Now, look up the chain manufacturer spec on #50 roller chain; You will find that it should be limited to 400 rpm and only 3 (three) horsepower (!). VERY conservative, YES?

GO KART CHAIN

Go kart racers run #35 pitch chain and they produce 18 horsepower at 18,000 rpm from a 6.1 cubic inch engine. Racing karts can exceed 150 miles per hour. The chain manufacturer lists that chain at 900 rpm maximum and 1.4 horsepower! Very conservative.

MOTORCYCLE BELTS

A new Harley Davidson motorcycle produces 65 horsepower and will cruise all day long on hot, desert highways and dusty gravel roads, with a Gates 14 mm x 37 mm wide poly chain belt driving the rear wheels. The Gates Catalog lists that belt at 23 horsepower at 870 rpm for a '28 tooth sprocket. That is very conservative also.

MORSE HY-VO CHAIN 4 WD TRANSFER CASES

Current production four wheel drive trucks and vans, with 350 cubic inch engines to 502 cubic inch engines, use a chain drive in the transfer case that measures 1.0" wide x 1/2" pitch. The catalog specifications for that chain show that it should be limited to 105 horsepower at 2,100 rpm. The 502 engine often produces over 500 horsepower at over 5,000 rpm. Again, the catalogs are extremely conservative.

WHAT DO WE BELIEVE?

Thankfully, all of the ground breaking inventions of aircraft chain drive units, gear drive units, and belt drive units believed what they saw in the real world of power transmission, and ignored the super conservative specifications in the manufacturers' catalogs. We can do the same thing. Remember the old adage that if you want to win a race, copy the previous winners. Don't copy the also-rans and the losers. In this chapter, you will be able to read about the specifications of the chains that have been proven in more than 20 years of successful flying.

SIZING THE CHAIN

Past performance has shown that a 1/2" pitch chain, 1 1/2" wide, will dependably handle the power from a 350 cubic inch V-8 auto engine. If you are a nut about charts, graphs and figures, you could plot your own horsepower versus chain width chart from that one known figure. The HY-VO chain catalog states that the 1/2" pitch, 1 1/2" wide chain has a static tensile strength of 15,000 pounds. You might calculate a 4" diameter drive sprocket (one third of a foot) would exact 1,500 foot pounds of torque on the chain at 6,000 rpm and 500 foot pounds of torque from the 502 cubic inch engine. There would be a lot of over-design in that size. This book will not attempt to resize and re-specify HY-VO chains

This cut-away of a GM T-400 trans axle from an Oldsmobile Aurora V-8 automobile, clearly shows the rather narrow Morse Hy-Vo chain that is expected to last the life of the car, in all kinds of driving conditions. And it really DOES last, too!

This 1.25 inch wide 1/2 inch pitch chain drives the front wheels in a Ford four wheel drive 1 ton truck. It easily lasts more than 250,000 miles without excessive wear. In low-low gear, the chain transmits over 1,500 foot pounds of torque, about 500% more than you need to turn a 3 blade prop on a V-8 engine.

CHAIN DRIVE CATALOGS
MORSE INDUSTRIAL
CATALOG PT - 88
Emerson Power
Transmission Corp.
Ithica, NY 14850

Here is an off-the-shelf Morse Hy-Vo chain and 2 sprockets that retail for $327.52. Compare that to the cost of 1 belt and 2 sprockets listed on pg. 84, at $423.56, and you may see that chain drives are the way to go. The machine shop work on belt drives is more simple than for splining the chain drive sprockets, but that expense will almost even out the total cost.

Perhaps the only disadvantage to chain drives compared to belt drives is the requirement for oil and an oil spray in the chain drive.

CHAIN DRIVE CATALOG
Ramsey Products Corp.
Catalog # 490
Design Manual # 840
P.O. Box 668827
Charlotte, NC 28266
1-704-394-0322

Steve Parkman, of SWAG Aeromotive, Tucson, Arizona, has designed this prototype 2 to 1 ratio roller chain drive for a 100 horsepower Geo auto engine. The chain and sprocket are from the cam timing on a 350 Chevy V-8 engine. Stacking two chains, one behind the other, would double the horsepower capability of the unit.

for auto engine use. This book will tell you what has worked in the past, and you can "copy the winner".

CHAIN WIDTHS THAT WORK

- 3/4" wide x 3/8" Pitch for 75 hp to 150 hp

- 1" wide x 3/8" Pitch for 150 hp to 250 hp

- 1 1/2" wide x 1/2" Pitch for 250 hp to 400 hp

- 2" wide x 1/2" Pitch for 400 hp to 600 hp

- 3" wide x 1/2" Pitch for 600 hp to 800 hp.

SPROCKET SELECTION

Remember that small diameter sprockets make the chain work harder than large diameter spockets do. That is because the chain is required to bend and straighten each time it passes around the sprocket at maximum operating speed. The smaller the sprocket, the tighter the chain bends/flexes and the faster it does this at any given rpm.

The smallest diameter Morse-type chain drive sprocket that will dependably transmit auto engine power to a propeller, is 3 1/2" diameter. For longer chain life, a 4.0" diameter drive sprocket will greatly increase the life of the chain.

It is easy to figure drive ratios when selecting sprockets. The driven sprocket will be exactly twice the diameter of the drive sprocket in a 2 to 1 ration. A 1/2" pitch, 25 tooth sprocket will be 3.989" diameter and in stock form, will weigh 4.9 pounds. A 1/2" pitch, 57 tooth sprocket will be 9.076" diameter and will weigh 27.1 pounds. You can turn the sprockets on a lathe to remove a lot of excess weight. The 4.9 pound drive sprocket can be turned down to

MAKING A PROTOTYPE CHAIN DRIVE UNIT

Designing and making a flyable chain drive PRSU is entirely possible on a small budget. The parts, bearings, and aluminum plates will cost less than $1,000. The machine shop labor is whatever you can bargain for. If you want to sign up for an evening class at your local trade school, you can do almost all of the of the machining yourself. However, you will need to "farm out" the gear and sprocket splining operations.

Very few trade schools are equipped to cut internal splines and external multiple splines. In cities of over 100,000 population, there are usually gear cutting shops that can cut the splines for you. For addresses, check the yellow pages of your phone book or go to a chamber of commerce office and ask them for their industrial phone book.

WOOD MOCKUP

The largest manufacturing companies in the world once started all new, complex designs, by making the part out of wood first. In more recent times, laser cutters are used to cut casting patterns out of thin paper or even thin aluminum sheets. Then, the laser cut patterns are glued together to form the prototype parts. You can still make your prototype out of blocks of wood or even sheets of 1/8" to 1/16" plywood.

TEST STAND

After you have made your prototype chain drive PRSU, it is highly advisable to test it on your engine, with a test club prop, for at least 1 week of 8 hour days. That test time will only be 40 hours or less. But it is better to bench test than to have a significant in-flight failure due to lack of testing.

It may well be that chain drive PSRU's will replace belt drive units in the next 5 to 10 years.

Investigate this design before trying other PSRU's.

SPRAY BARS & OIL PUMPS

We found, in testing chain drives for the auto manufacturers, that ATF fluid (automatic transmission fluid), worked best for chain lubrication. You can incorporate a tiny oil pump from a 4-wheel drive transfer case to spray oil onto your chain and sprockets. You do not need to use engine oil for the chain drive system. Take a look at a late model Jeep transfer case for a good oil pump design that is part of the lower input shaft.

---

weigh 3 pounds, and the 27.1 pound drive sprocket can be turned down and drilled to weigh 8 pounds or less. See the sketch in this chapter for possible weight removal.

WEIGHT REMOVAL

Stock, off the shelf Morse chain sprockets are very heavy, partly to save machining costs and partly to allow the transfer case designer the opportunity to selectively remove weight. The stock sprockets are usually 4 to 5 times heavier than is required for auto engine transfer case use.

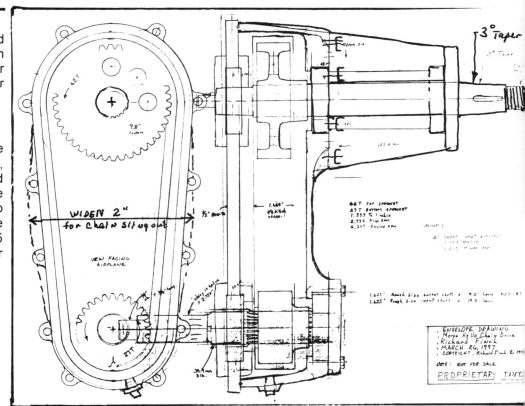

This is a rough sketch of a Morse Chain PSRU that incorporates the chain and sprockets shown on page 95. Milling the housing out of 7075-T7 or 6061-T6 aluminum would work just fine. Aluminum flat stock comes in thicknesses of 1" up to 12".

Gear Box Drive Designs

◆◆

I designed this 1.63 to 1 ratio gear box for auto engines of 200 to 350 horsepower. This gearbox PSRU is obviously a prototype, but the 1 1/4 inch wide gears will handle 800 horsepower in a drag boat, so gear dependability should not be a problem. The unit weighs 63 lbs. It could be 20 lbs. lighter and still handle 250 horsepower.

PROTOTYPE COSTS

The 1.63 to 1 ratio PSRU gearbox shown on this page was supposed to cost $995. Before it was completed by the boat gear box machine company, it cost me $2,000. It also came to me weighing 63 pounds, 20 pounds more than the design called for.

The problem with the cost and the weight was caused by the "dreadful word", AIRPLANE ! Once the owner of the boat drive machine shop realized this gearbox might actually fly, he became airplane oriented. This means money and lawsuits to a lot of people

If you decide to follow this example, don't say "AIRPLANE" at any time. Call it an "off-road transfer case" or a "racing boat gear box". Most uninformed people believe that airplanes will all "crash and burn", especially the experimental planes.

Pictured on the top of this page is an engineering drawing of a prototype gearbox PSRU for a big block V-8 engine, expected to produce over 500 horsepower. Note in the drawing that a significant flywheel is incorporated and that a large elastonomeric drive bushing is used to cut down on vibration and gear shock loads.

The brochure on this drive unit advertises that the unit housing is fully machined from aluminum billet to reduce development costs and weight.

By reducing the width and weight of the gears, and by reducing the thickness of the housing material, the weight can be pared down to about half of the 90 pounds specified. If you only need to transmit 200 to 250 horsepower, and not 500 horsepower, you can do with narrower gears and thinner metal in the housing.

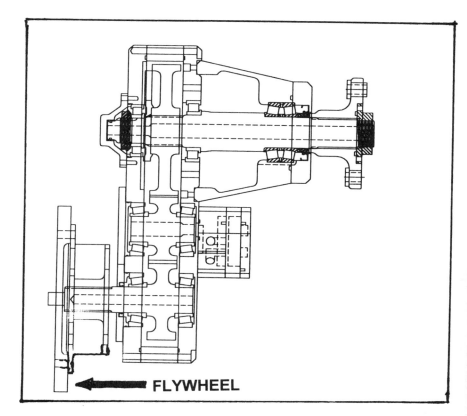

FLYWHEEL

Above is a development drawing of a 500 horsepower gear box to be milled from solid aluminum billet stock. Finished weight is 90 pounds. A 250 horsepower unit could safely weigh 45 pounds. Drawing: CAM-FIRE Engineering, Sidney, B.C., Canada

THE FINCHBIRD PSRU

When I paid for and picked up the gearbox PSRU (at the machine shop) I asked the proprietor how much horsepower this unit will handle if it were installed in a racing boat. His answer was, "Oh, it should handle 800 horsepower or more in a nitro burning drag boat!" (I had specified 200 to 250 horsepower maximum in my purchase order!)

What this means to you is that people in the automotive modification business simply cannot think in any terms except producing the maximum power possible and beating everyone else to the finish line. They can't think of reduced power and light duty use.

Without a single exception, I have never discussed engine and gearbox use in airplanes with race car people who don't fixate on 9,000 rpm, 1,000 horsepower, and winning any race that the engine might have the chance to run in.

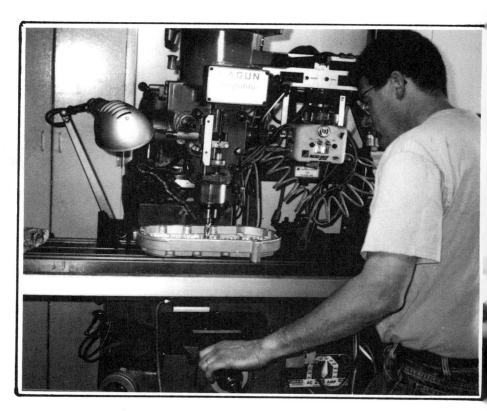

Dale Johnson uses his home workshop Lagun mill to remove excess weight from my FINCHBIRD gearbox casting. We removed over 10 pounds of unnecessary weight from the casting.

In this photo, I am welding up a hole in my gearbox casting where I machined away too much metal. The only accurate way to weld thick castings is to TIG weld and to pre-heat the casting to 350° F for one hour before welding.

Here is the FINCHBIRD gear box with all the gears installed, ready to press the second half of the housing onto the gear. The top gear is the prop shaft, the center gear is the idler gear, and the bottom gear is the engine crankshaft input.

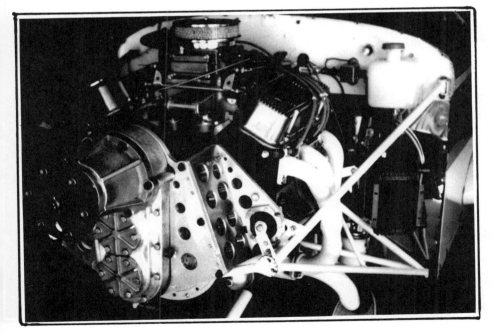

Here is the gearbox, filled with oil, test run, and ready to bolt the prop onto. Design improvements in this prototype would include shortening the fore-to-aft mounting plates and making a 2 piece rather than 3 piece housing.

The moral to this part of this story is, that if you discuss airplane engines with race car people, expect to accept sky-high horsepower, overhauls every 10 to 15 hours, and terribly inflated prices because "airplanes crash and you get sued!"

COST OF GEAR DRIVES

Check the small ad on page 104 to see how inexpensive a 250 hp to 450 hp race car gear box can be. If they can sell them for $995.⁰⁰, airplane gear boxes COULD be that cheap also. Richmond Gear makes and sells this gearbox.

George Morse uses second gear from a MUNCIE truck transmission to get a 1.66 to 1 ratio in his very compact and light weight PSRU gearbox. This unit has been dependable in test stand runs and in over 900 hours of flight time, since 1974 or earlier.

A very unique and very well tested and proven gear box design is this 2 gear unit from George Morse's Prowler airplane engine. It is mounted on the FRONT of the 215 Buick V-8 engine, rather than the rear, therefore 2 gears will turn the prop in the clockwise rotation.

2 GEAR PSRU'S

When you use two gears to make a PSRU, you reverse the normal rotation of the engine. If the flywheel of the engine turns clockwise when it operates, then the propeller will turn counter-clockwise. It is harder to find factory built counter-clockwise propellers, so you really should prefer a clock-wise prop rotation.

There is an easy solution to this dilemma. You just turn the engine around and drive the prop off the V-belt pulley end of the engine. Since you need to make a special gearbox housing anyway, you just make the gearbox bolt pattern so that it attaches to the front of the engine rather than the back of the engine.

GEORGE MORSE GEARBOX

In as early as 1975, George Morse, then living in Soqual, California (the Monterey area), designed the very dependable 2-gear PSRU shown on this page. The story goes that he used second gear from a GM truck transmission because it was a 1.66 to 1 ratio. In the photos above, you can see the bevel cut gears that are about 3/4" wide, only half as wide as the gears in the prototype gearbox pictured on page 99.

This excellent gearbox provides oil pressure to operate a certified constant speed propeller. Plus, it appears to weigh about 30 pounds. That gearbox has performed flawlessly for over 20 years and for over 1,000 hours total. If you need less than 200 horsepower from your V-6 or 4 cylinder engine, you might investigate the possible use of automobile transmission gears and bearings. Another feature of this PSRU design is that the flywheel end of the engine is located toward the firewall. This engine uses a somewhat heavy flywheel to make up for deleting the crankshaft harmonic balancer.

Everett Hatch has designed and built this exceptionally clean __INTERNAL__ gear PSRU for his Mazda rotary engine in his RV-4 airplane. Internal gear designs do not reverse the output rotation of an engine as external gears would do, and internal gears keep more teeth in mesh than external gears do. See the sketch below for how internal gears work.

This Ross planetary gearbox on a Subaru Legacy engine keeps the crankshaft and prop center lines on the same plane, but requires a reverse, counter-clockwise prop because this particular planetary system reverses the input and output rotations.

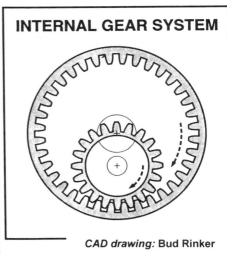

INTERNAL GEAR SYSTEM

CAD drawing: Bud Rinker

This sketch shows how the Everett Hatch internal gear PSRU keeps the prop rotation the same as the engine rotation. Ninety-nine percent of all auto engines turn clockwise when you are facing the water pump/pulley end of the engines.

The business end of a Subaru Legacy 2.2 liter engine shows that in-line PSRU's or off-set PSRU's will work equally well on this 265 lb. all aluminum engine.

This is a close up picture of a Bud Rinker VW gearbox and prop hub on a Corvair engine. I converted one of Bud Rinker's designs to fly on a Cessna 150 airplane. The Corvair engine is the 1% of most auto engines. It turns counter-clockwise when facing the pulley end of the engine, making a 2-gear PSRU work perfectly for standard airplane propellers.

PLANETARY PSRU'S

Opposed cylinder design engines such as Corvair's, Subaru's and rotary engines such as the MAZDA 13B engine work very well with planetary gear boxes.

Planetary gears come from automotive automatic transmissions. The planetary gears allow the input and output shafts to be placed in the same concentric plane. One problem with planetary gear boxes is that they usually (but not always) reverse the rotation of the output shaft and you must use a reverse, counter-clockwise propeller. Planetary gear PSRU units have been used for many years, even on certified aircraft engines.

VERTICAL SYSTEMS

Check out the addresses in Chapter 19, for Lars Nelson and his Corvair V W gear box drive units for the Corvair engine. Bud Rinker, of Santa Barbara, CA, designed and tested the Rinker drive units on Corvair engines in the 1960's.

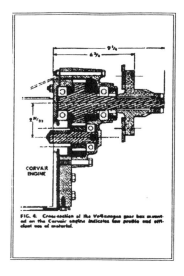

Bud Rinker's sketches of his VW gearbox adaptation to a 150 horsepower turbocharged Corvair engine. Plans are sold by Lars Nelson, Vertical Systems, Santa Barbara, California

RINKER PSRU

Pictured below on this page, are the specification drawings for a plans-built gear drive unit for 80 to 180 horsepower Corvair engines. Bud Rinker, of Santa Barbara, California, designed the PSRU in the early 1960's. The plans are now sold by Lars Nelson of Santa Barbara, operating Vertical Systems.

An excellent design for a lower horsepower, 2 cycle, ultra light engine gear box; the large gear on the right is the helical driven gear on the prop shaft, and the small helical gear in the housing on the left is the engine crankshaft drive gear. Helical gears, cut at an angle, are much quieter than spur, straight cut gears, but helical gears also exert significantly more thrust on the bearings than straight gears do.

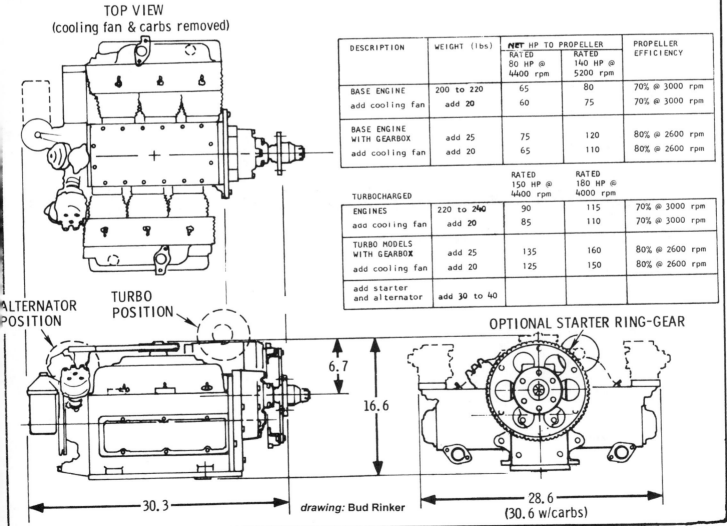

TOP VIEW
(cooling fan & carbs removed)

DESCRIPTION	WEIGHT (lbs)	NET HP TO PROPELLER		PROPELLER EFFICIENCY
		RATED 80 HP @ 4400 rpm	RATED 140 HP @ 5200 rpm	
BASE ENGINE	200 to 220	65	80	70% @ 3000 rpm
add cooling fan	add 20	60	75	70% @ 3000 rpm
BASE ENGINE WITH GEARBOX	add 25	75	120	80% @ 2600 rpm
add cooling fan	add 20	65	110	80% @ 2600 rpm

		RATED 150 HP @ 4400 rpm	RATED 180 HP @ 4000 rpm	
TURBOCHARGED ENGINES	220 to 240	90	115	70% @ 3000 rpm
add cooling fan	add 20	85	110	70% @ 3000 rpm
TURBO MODELS WITH GEARBOX	add 25	135	160	80% @ 2600 rpm
add cooling fan	add 20	125	150	80% @ 2600 rpm
add starter and alternator	add 30 to 40			

ALTERNATOR POSITION

TURBO POSITION

OPTIONAL STARTER RING-GEAR

6.7

16.6

30.3

drawing: Bud Rinker

28.6
(30.6 w/carbs)

SPUR GEARS

Gear design engineers must consider many factors when designing a gear transmission for a specific use. Spur Gears, being straight-cut, do not exert any significant thrust loads on the bearings. Helical gears, while quieter and smoother in operation, can exert 50% or more torque thrust into the shaft bearings. Therefore, as the horsepower requirements go up, the advantages of straight cut spur gears increases. In the S-51 Stewart P-51 Mustang replica, the big-block Chevy engine produces so much torque and horsepower that spur gears make good sense.

One major disadvantage to spur gears is that they "whine" in normal operation. In a P-51 replica, the "whine" sounds natural, so it can be a good thing. Helical gears do not whine because they "slide" each gear tooth into engagement.

Total Engine Concepts, of Florida, builds this spur gear PSRU for the big block Chevrolet V-8 engine used in the Stewart S-51 Mustang replica. The housing is not shown in the photo.

Check out the price on the 2 speed gear box for race cars. Airplane gear boxes COULD be this price also.

The two massive spur gears shown here are not as heavy as they look, because a lot of weight has been removed from the inside of the gear without sacrificing strength. These gears are for the Stewart S-51 replica fighter and easily handle 500 or more horsepower.

Pictured to the right, on page 105, is a side view drawing 1/5th scale, of a Chevrolet V-6 engine. You can use this drawing to lay out a plan for a Chevy V-6 engine in your airplane. The gearbox design sketched at the flywheel end of the engine is for sizing only. The size of the gearbox shown would also work for a chain drive or for a cog belt drive.

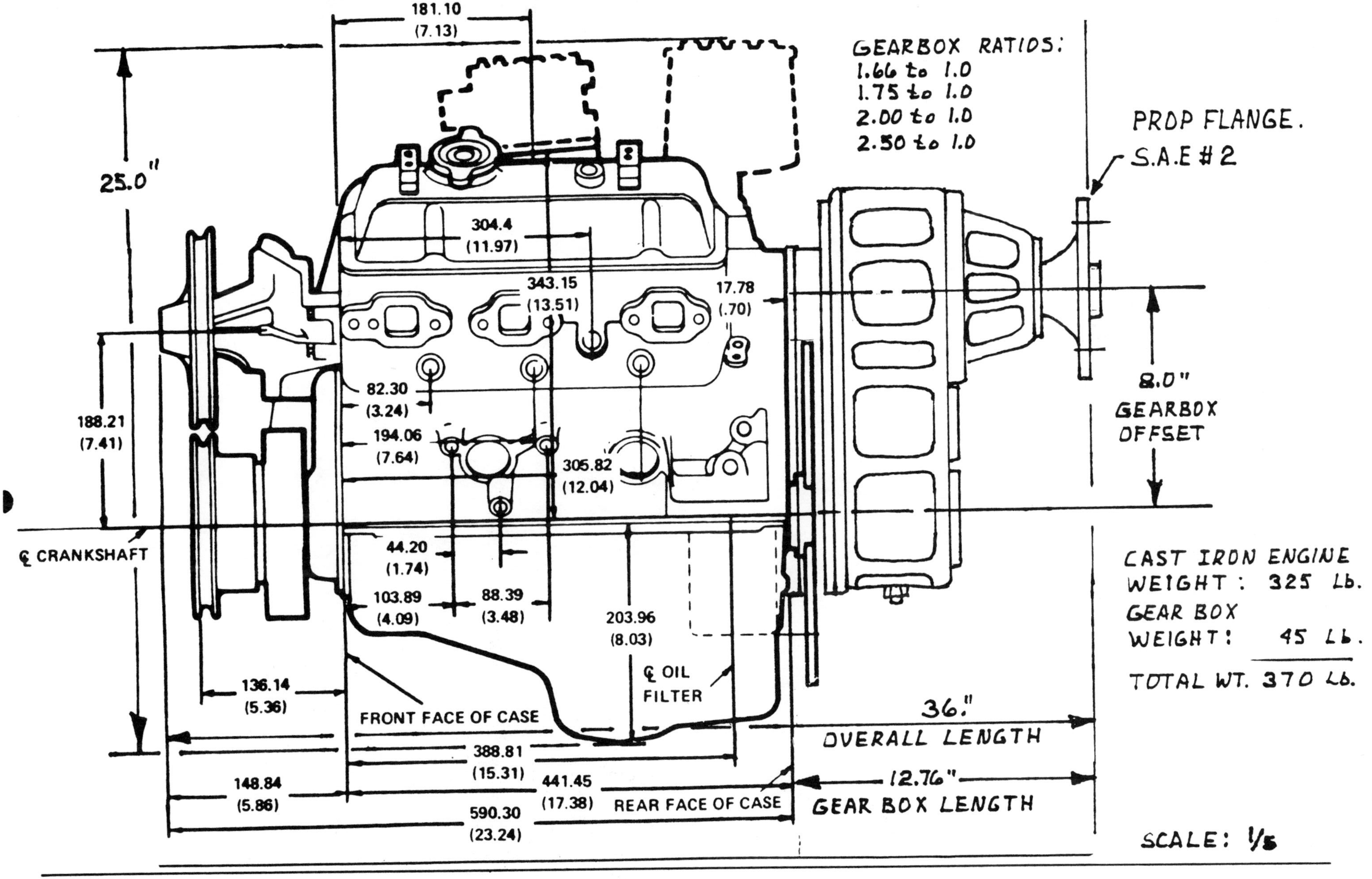

PROP FLANGE.
S.A.E #2

8.0" GEARBOX OFFSET

GEARBOX RATIOS:
1.66 to 1.0
1.75 to 1.0
2.00 to 1.0
2.50 to 1.0

CAST IRON ENGINE
WEIGHT: 325 Lb.
GEAR BOX
WEIGHT: 45 Lb.
TOTAL WT. 370 Lb.

36"
OVERALL LENGTH

12.76"
GEAR BOX LENGTH

REAR FACE OF CASE

17.78
(.70)

305.82
(12.04)

343.15
(13.51)

304.4
(11.97)

203.96
(8.03)

℄ OIL
FILTER

88.39
(3.48)

82.30
(3.24)

194.06
(7.64)

44.20
(1.74)

103.89
(4.09)

FRONT FACE OF CASE

388.81
(15.31)

441.45
(17.38)

590.30
(23.24)

181.10
(7.13)

136.14
(5.36)

148.84
(5.86)

25.0"

188.21
(7.41)

℄ CRANKSHAFT

CHEVROLET 90° V-6 ENGINE
DIMENSIONS & PROPOSED GEARBOX
© 1991 FINCH BOOKS

SCALE: 1/5

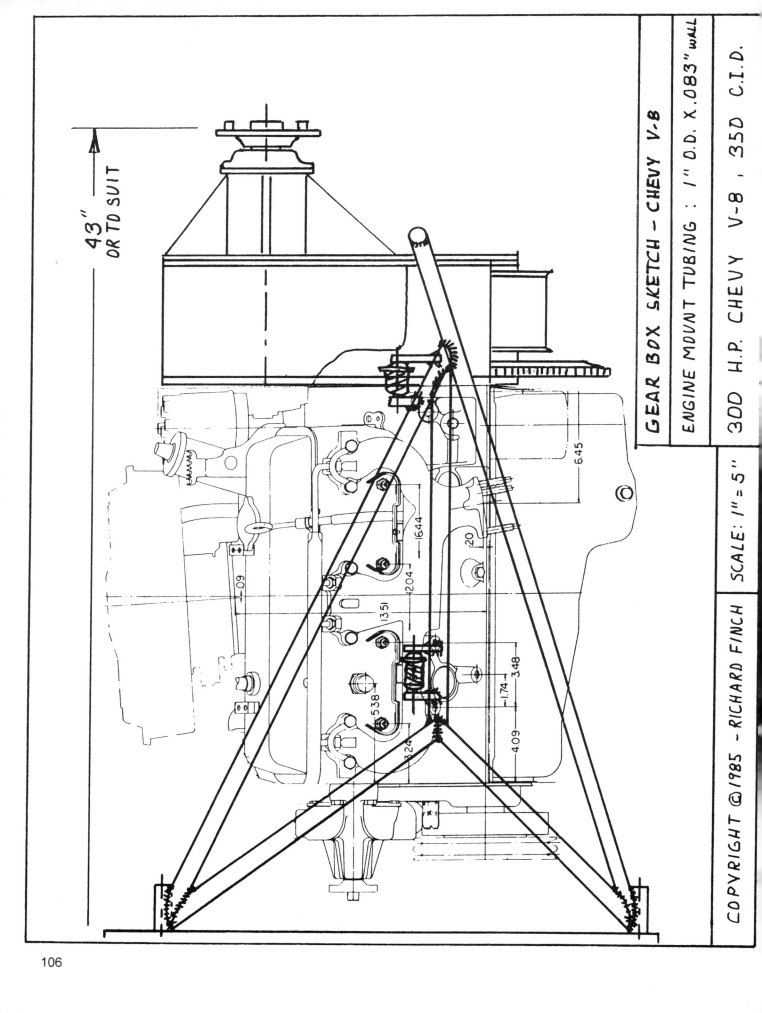

GEAR BOX SKETCH - CHEVY V-8

ENGINE MOUNT TUBING : 1" O.D. X .083" WALL

300 H.P. CHEVY V-8 , 350 C.I.D.

43"
OR TO SUIT

6.45

16.44

.20

12.04

13.51

1.09

5.38

3.24

1.74

3.48

4.09

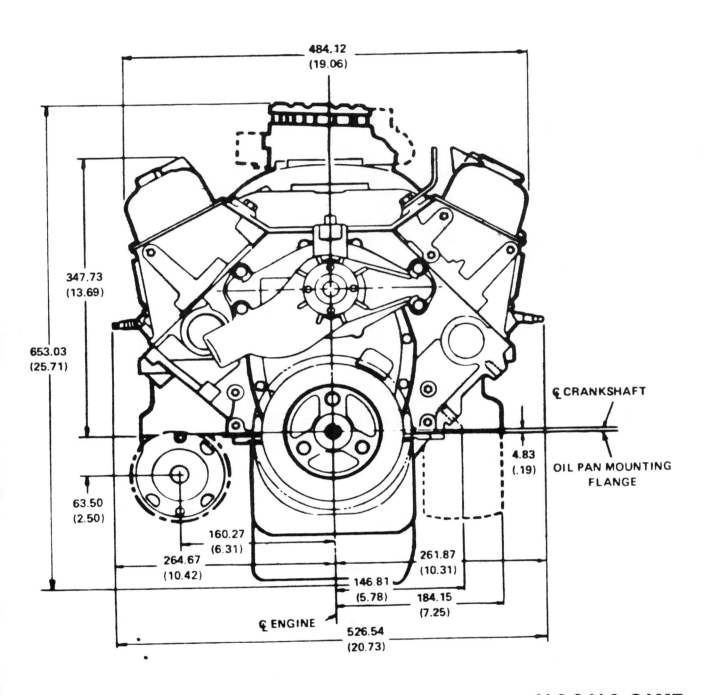

484.12
(19.06)

347.73
(13.69)

653.03
(25.71)

¢ CRANKSHAFT

4.83
(.19)

OIL PAN MOUNTING
FLANGE

63.50
(2.50)

160.27
(6.31)

264.67
(10.42)

261.87
(10.31)

146.81
(5.78)

184.15
(7.25)

¢ ENGINE

526.54
(20.73)

V-6 & V-8, SAME

FRONT VIEW - 1/5 SCALE

CHEVROLET 90° V-6 ENGINE DIMENSIONS FOR ENGINE CONVERSION FOR AIRCRAFT USE.

CHEVROLET 90° V-8 ENGINE WIDTH IS SAME AS V-6.

Some very good design ideas are shown here in this Rolls Royce Merlin P-51 and P-38 gearbox housing. If you decide to make castings for your gearbox, copy the bolt pattern that you see here in this photo.

ALUMINUM FOUNDRIES

Aircraft Foundry Co. Inc.
5316 Pacific Blvd.
Hunington, CA 90255
(213) 587-3171

**American International
Engineering and
Manufacturing Inc.**
555 Spring Road
Moorepark, CA 93021
(805) 529-1488

Future Aluminum Factory
6107 Wilmington Ave.
Los Angeles, CA 90001
(213) 589-1861

Shell Pattern & Foundry, Inc.
1802 Glassell
Orange, CA
(714) 637-5343

GEAR MANUFACTURERS

Gear Manufacturing Inc.
1325 Red Gum Avenue, Suite 17
Anaheim, CA 92806
(714) 666-2611

Gerhardt Gear Co. Inc.
3060 No. California
Burbank, CA 91504
(818) 842-6700

Fisher's Gear & Machine Co.
1201 So. Santa Fe
Los Angeles, CA 90021
(213) 624-7554

Globe Gear
370 Commerce Dr.
Fort Washington, PA 19034
(800) 523-2576

Direct Drive Designs

◆◆◆◆◆◆◆◆◆◆◆◆◆◆◆◆◆◆◆◆◆◆◆◆◆◆◆◆◆◆◆◆◆◆◆◆

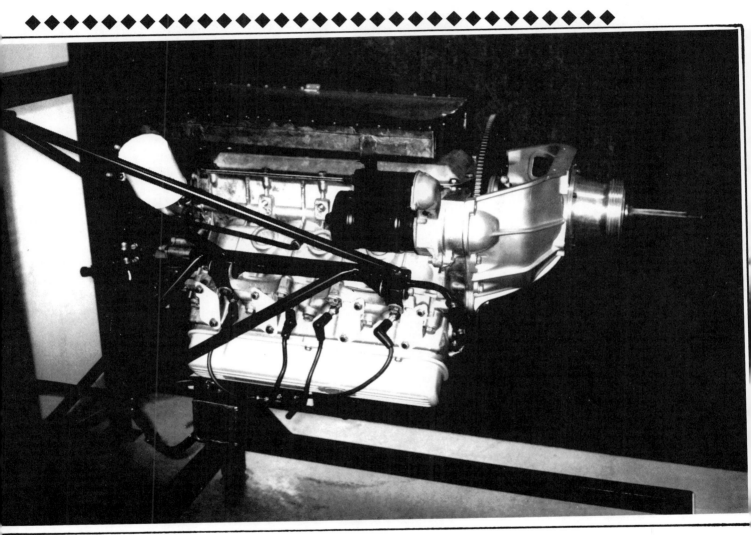

The Steve Wittman invented Buick V-8 engine shown on this display stand, was the example that many people followed in converting auto engines for experimental aircraft. By studying this photo carefully, you can see the many extremely simple features that made this conversion work flawlessly for over thirty years:
* The prop shaft is supported against gyroscopic forces by using a stock clutch/bell housing from a stick shift engine.
* The radiator, custom made from brass, lays flat on the flat oil pan area of the engine and the air enters the cowling, flows over the engine block and out the top of the cowl through louvers.
* The remote oil sump, a square box under the valve covers, collects 5 quarts of oil that drains from hoses exiting the valve covers, and is picked up by the external Buick oil pump via a rubber suction hose in the oil sump.

WITTMAN V-8

Steve Wittman converted the inverted Buick-Olds-Pontiac V-8 engine to direct drive in the mid-1960's, because cog belt drive units, PSRU's, were almost unheard of at that time. Even today, a pre-fabricated PSRU for the B.O.P. V-8 engine will cost over $3,000, and yet the V-8 engine can be purchased for $500 and overhauled for another $500 or less.

The 215 cubic inch, 3.5 liter V-8 engine was rated at 200 horsepower at 5,000 rpm, with 11.0 to 1 compression. With a more realistic 9.0 to 1 compression ratio, the 215 aluminum V-8 was rated at 155 horsepower at 4,600 rpm.

In flight test reports done by several aviation writers, comparing the B-O-P V-8 to Wittman Tailwind airplanes powered by 125 horsepower Lycoming engines, the flight tests reported 125 to 135 horsepower available from the V-8 direct drive engine.

EASIER CONVERSION?

To most people who are starting to investigate the potential benefits of converting to auto engine power for their airplanes, it seems that simply bolting a propeller to the auto engine crankshaft would save a lot of time, worry, and especially, money.

And these people are surely right. That is, if they don't have to worry about forfeiting about 40% of the auto engine's available horsepower. In some cases they may forfeit even more than 40% of the power available. Take a close look at the horsepower chart on page 21. If you can turn your smaller-than-average diameter propeller to 3,000 rpm, you will get 85 horsepower out of the 2.3 liter, 160 horsepower Oldsmobile Quad 4 engine.

But if you can only turn your prop to 2,600 rpm, you will only get 70 horsepower out of the 160 horsepower engine. That is a 64% loss of horsepower.

Chris Beachner flew his NON-inverted Buick V-8 with a Wittman-type prop shaft, for over 1,000 hours, but at 135 to 140 horsepower from the 200 horsepower engine.

CHRIS BEACHNER V-8

In the late 1970's, Chris Beachner of Eloy, Arizona, built a neat looking 2 place tandem aircraft and installed a 215 cubic inch Buick V-8 engine in it, except that he mounted the engine right-side-up, not inverted. He used a 64" to 68" diameter wood prop, and turned the engine to 3,600 rpm on take-off and on full power high speed passes at air shows.

Several years into the test program on this simple and inexpensive auto engine conversion, Chris Beachner installed a small belt driven supercharger on the Buick V-8 engine and reported a 20% increase at full power. He was getting about 135 horsepower out of the V-8 engine prior to installing the supercharger.

Then, the last few months of the flight test program, Chris Beachner designed and installed a cog belt PSRU, and he told me at Oshkosh that year (1985 or 1986) that he SHOULD have converted to cog belt drive at the beginning of his flight test program. He said that he gained back all the lost horsepower plus a little extra, due to the new PSRU and to the belt drive supercharger. Sadly, Chris Beachner died in a post-crash fire when the engine died on take off from an airport in Washington state.

Chris had over 1,000 hours of trouble-free flight time on his Buick V-8 engine at the time of his death. The cause of the hard landing was believed to be a failure of a new set of ignition points that had just been installed minutes before the last flight. Possibly, an adjustment screw had worked loose.

———   ———   ———

HORSEPOWER AT LOW RPM

Here is another comment on that subject. Understand that any piston, internal combustion engine, is primarily an air pump. The more air (and 14.0 to 1 fuel) it can pump in one minute, the more horsepower it can produce by the fuel combustion process. Most modern engines can produce 1 horsepower per cubic inch of engine displacement when turning 5,250 rpm. The only way to produce 1 hp per cubic inch displacement at 2/3 of the 5,250 rpm, is to pack in more air and fuel by either turbocharging or by supercharging. Re-grinding the camshaft just can't do it. Read page 115 for further information about Turbocharging and supercharging.

Don Sauser, of Tustin, California, has done an excellent job of engineering a "crate engine" 350 Chevrolet V-8 engine in his P-6E replica. Study this photo for good engine mount ideas.

DON SAUSER CHEV V-8

Take a close look at the three photos of Don's beautiful replica Curtiss P-6 E. Hawk WWI trainer airplane that is scaled 82% and powered by a 350 C.I.D., 345 horsepower small block Chevrolet V-8 engine with aluminum cylinder heads. The engine cost <u>new</u>, $2,500, NOT $25,000!

Even though the 82% scale airplane needed only 180 horsepower to fly, Don found that a little more horsepower would be nice to have. Turning the wooden prop, directly mounted to the engine, likely at less than 1/2 of the rated rpm of 5,200 rpm, produces less than half of the engine's rated horsepower, or about 160 horsepower. In a very sucessfull effort to maintain an accurate scale appearance, a log type exhaust manifold with 12 short stacks on each side, for a total of 24 exhaust stacks, was used. In this exhaust system, all 24 stacks puff smoke and exhaust noise for super authenticity. But a race car type header pipe with 22 dummy exhausts stacks would add about 10 to 20 more horsepower to the engine at 2,400 rpm or less.

If you happen to see Don Sauser's P-6E at a fly-in, take a close look at the airplane. It is a real work of art, and Don designed it and built it in his garage at home. The first engine runs were done in his driveway, with understanding neighbors looking on.

Be sure to pay particular attention to details on Don's P-6E, in his radiator design. You can see in these pictures that the radiator is mounted completely below the level of the engine. The reason for this is to maintain a completely authentic appearance. The original P-6E had its radiator in the same location.'

You can also see that the Chevrolet V-8 engine is equipped with dual coil ignition, similar to the Jess Meyers and Jeff Ackland designs.

This right-side view of Don Sauser's P-6E replica shows the wooden 2-blade prop bolted to the Chevy V-8 crankshaft via an adapter.
Don Sauser Photo

THE REASONS FOR DIRECT DRIVES

In the case of the two airplanes pictured on this page, authenticity is the first reason to keep direct drive propellers. Both airplanes are considered to be classic designs and avoiding extra lumps and bumps from PSRU's is important in keeping these designs looking authentic. The second and very important reason for designing direct drives, is the desire to avoid the extra cost and weight of a PSRU. Before you make the final decision about to build or not to build or buy a PSRU, talk to both Don Sauser and William Wynne about these two direct drive designs.

It is not easy to make a replica airplane look EXACTLY like the real thing, but Don Sauser has surely succeeded in this direct drive Chevrolet V-8 conversion for his 82% Curtiss P-6E replica. Don Sauser Photo

The partners at Precision Arial Services, in Daytona Beach, Florida, have installed a 100 Horsepower direct drive Corvair engine in this Cassutt Racer that was originally designed for a 100 horsepower Continental 0-200 engine. The weights come out the same, but the Corvair SOUNDS much better than the Continental 4 cylinder.

PRECISION AERIAL SERVICES CORVAIR

Since the first six cylinder Corvair engine first came on the market in October 1959, airplane people have been intrigued with its aircraft engine appearance and extremely low cost. In 1979, you could still buy factory new Corvair engines in crates from your local Chevrolet dealer for less than $1,500.

Today, you can buy used Corvair engines for $50 to $400, depending on the condition and on the rated horsepower. Corvair engines were made in 80 horsepower to 180 horsepower, with a 95 horsepower version rated for full power at 3,600 rpm.

However, if you want to take full advantage of a Corvair engine's full potential, it still pays to incorporate a cog belt drive PSRU or a Bud Rinker gear drive PSRU.

A big advantage of the 164 cubic inch, 2.7 liter 6 cylinder air cooled engine, is that it only weighs 200 pounds, bare. Compare that weight with a modern, water cooled, all aluminum PVR, (Peugeot, Volvo, Renault) 2.7 liter, 164 cubic inch engine that weighs 90 pounds more, at 290 pounds. For more information about Corvair engines for aircraft, read my book *How To Keep Your Corvair Alive* .

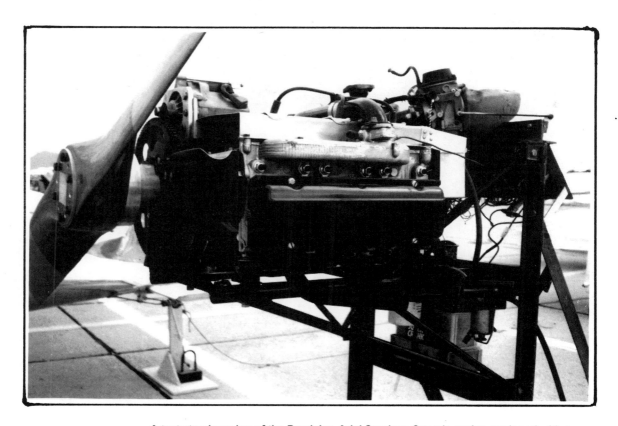

A test stand version of the Precision Arial Services Corvair engine equipped with a NISSAN Sentra flywheel, a Toyota Tercel starter, and a Kubota or garden tractor alternator. Total weight is 222 pounds, with direct drive, and producing110 horsepower.
Photo taken at the Copperstate Fly-In, 1997.

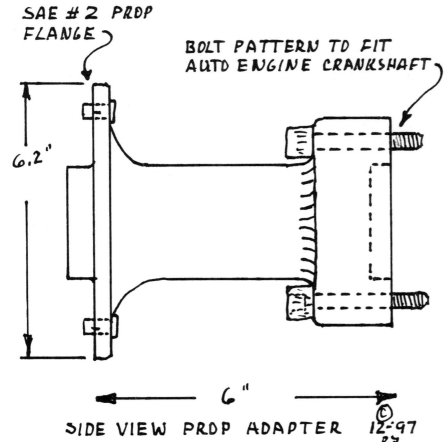

SAE # 2 PROP
FLANGE

6.2"

BOLT PATTERN TO FIT
AUTO ENGINE CRANKSHAFT

6"

SIDE VIEW PROP ADAPTER   12-97
MATERIAL: 7075-T7 ALUMINUM.   R.7.

ADAPTING THE PROPELLER

Simply bolting a propeller to the flywheel end of an auto engine crankshaft, is not possible. You must still make an adapter to connect the prop to the engine.

The most simple adapter is an aluminum spacer that is made from a piece of 2024-T4, 6061-T6, or 7075-T7 aluminum bar stock. The sketch on the left shows one possible suggestion for a prop extension/drive hub adapter for a forged steel crankshaft V-8 engine.

In two successful direct drive airplanes, the Wittman Tailwind and the Beachner Special, those airplanes used a bell housing and a special prop shaft extension. See page 109 for a picture of the Wittman Direct Drive.

DIRECT DRIVE PONTIAC V-8

The trailer mounted, test stand, V-8 engine shown in this chapter, is an all cast iron, 425 cubic inch engine that develops 370 horsepower at 4,800 rpm. At 2,600 rpm, it develops roughly half of its rated power, about 180 horsepower.

This direct drive Pontiac V-8 engine is set up on a test stand that can measure torque while the engine is running, and therefore output horsepower can be calculated. The screen around the 3-blade prop protects workers and onlookers from the dangers of the turning prop.

RE-GRINDING THE CAMSHAFT

For many years, I have heard people say they were going to re-grind the engine's camshaft to lower the rpm that the engine produced its horsepower at. Realistically, if that were possible, the auto engine manufacturer would have done that from the beginning.

You can raise the rpm and horsepower of a stock auto engine, by changing the camshaft, the exhaust system, the intake system, and by porting the engine for better breathing, but, sorry, you can't make an engine develop maximum power at 2,600 to 2,700 rpm, by simply re-grinding the camshaft.

RESIST DIRECT DRIVES

The moral to this chapter is that you should only plan a direct drive conversion for your airplane if you can accept just half the horsepower that your engine is capable of producing.

And before you make that important decision, take a 4-speed transmission car for a drive, using ONLY THIRD GEAR. You will quickly see that take-off's and hill climbing are not very practical when your car is in the wrong gear.

TAYLOR BIRD SUBARU AIRPLANE

Mr. G.C. Taylor, the designer and builder of the Taylorcraft line of certified airplanes, designed and built this Subaru pusher cog belt drive design in the early 1970's. (The first known use of cog belt drive PSRU's was Waldo Waterman in the 1960's. Mr. Waterman used Chevrolet Corvair engines in pusher design airplanes to test the concept of belt drive reduction units.) Mr. G.C. Taylor was the second known person to design and fly a cog belt drive reduction unit. Mr. Don Taylor (no relation) was the third person to fly cog belt reduction drives.

Reg Clarke, Operations Manager of Air-Ryder Manufacturing, Inc., Wetaskiwin, AB, Canada, has flown this Subaru EA-81 powered Dragonfly airplane for more than 500 hours with the direct drive engine. To offset the loss of power at the lower direct drive rpm, Reg has installed a turbocharger to boost the engine performance back up to 100 horsepower.

DIRECT DRIVE AND TURBOCHARGING

During the past 50+ years that I have been interested in piston engines, I have learned that it is not wise to say "It can't be done". Just about the time you say that, some ingenious engineer and tinkerer comes along and "does it".

But to this date I can truthfully say that "IT" has not been done YET. That is, no one has managed to re-grind a camshaft to produce maximum horsepower at 1/2 to 2/3 of the rpm that the engine can produce at 5,250 rpm. Normally aspirated, that is.

If you turbocharge or supercharge the engine, you can pack in nearly as much air and fuel at 2,750 rpm, as you can get into the engine at 5,250 rpm. And it takes a certain number of molecules (pounds) of fuel and air to produce one hundred horsepower in a normally aspirated internal combustion engine.

For the sake of this discussion, let us say that a 500 cubic inch engine, turning at 5,250 rpm, can draw in 1,312,500 cubic inches of air and fuel in one minute, or 759.5 cubic feet of air and fuel per minute. When that fuel burns, and pushes the pistons down to rotate the crankshaft, the engine can produce 500 horsepower. Now, if we reduce the rpm to 2,750, the 500 cubic inch engine will only draw in 687,500 cubic inches of air/fuel in one minute, or 397.8 cubic ft. of air/fuel in one minute.

We already found that each cubic foot of air/fuel produces about .658 horsepower. That means that our 500 cubic inch engine, turning at 2,750 rpm, will produce 261.7 horsepower, not 500 as before. Strangely enough, that is what our torque and horsepower chart verifies.

But, if you compress the air/fuel inlet charge by turbocharging or by supercharging, you can often pack in as much as 759.5 cubic feet of air/fuel mixture at the lower rpm of 2,750. On paper, and in real life, you can get your 500 horsepower back at a lower rpm. But you can also get 50% more horsepower at 5,250 rpm. Meaning that now you can get 750 horsepower from your 500 cubic inch engine if you boost it from 29.92 " HG to 40" HG manifold pressure. Read more in the Richard Finch book, "TURBOCHARGERS".

WEIGHTS, H.P., C.I.D. SPECS OF AIRCRAFT ENGINES

AIRCRAFT ENGINE	YEAR	ENG. TYPE	LITERS	CU. IN.	HORSEPOWER @RPM	TORQUE	WEIGHT
Continental							
C-0200-A	1983	OHV F-4	3.3	200	100@2750	–	220
C-IO360G,N	1983	OHV F-6	5.9	360	210@2800	–	357
C-TSIO 360-C	1983	OHV F-6T	5.9	360	225@2800	–	320*
C-IO470-L	1983	OHV F-6	7.7	470	260@2625	–	493
C-IO520A,B.C	1983	OHV F-6	8.5	520	285@2700	–	469
C-IO520-E	1983	OHV F-6	8.5	520	300@2700	–	495
C-TSIO520-AF	1983	OHV F-6T	8.5	520	285@2600	–	456*
C-GTSIO-520-C	1983	OHV F-6TG	8.5	520	340@3200	–	567*
C-GTSIO-520-K	1983	OHV F-6TG	8.5	520	435@3400	–	619*
Lycoming							
L-0235-L2A	1983	OHV F-4	3.9	235	118@2800	–	242
L-0320-A2B	1983	OHV F-4	5.2	320	150@2700	–	264
L-0320-D	1983	OHV F-4	5.2	320	160@2700	–	275
L-IO360-A1A	1983	OHV F-4	5.9	360	200@2700	–	313
L-IO360-B1F	1983	OHV F-4	5.9	360	180@2700	–	290
L-TO360-E1A6	1983	OHV F-4T	5.9	360	180@2575	–	335
L-IO540-K1A5	1983	OHV F-6	8.9	540	300@2700	–	463
L-IO720-A1B	1983	OHV F-8	11.8	720	400@2650	–	586
Thunder Engines							
TE495-TC 700	1983	OHV V-8TG	8.1	495	700@4400	–	712

NOTE: In the Engine Type Column, OHV-means overhead valve

NOTE: Most airplane engine weights do not include starter, alternator, and baffeling. Add 40 to 100 pounds to the weights shown in this chart for true weights.

F-means flat or opposed cyl.
T-means turbocharged
G-means geared
*-means wt.(probably does not include turbocharger or waste gates.)

Propeller Choices

◆◆◆◆◆◆◆◆◆◆◆◆◆◆◆◆◆◆◆◆◆◆◆◆◆◆◆◆◆◆◆◆◆◆◆◆

The MT Propeller Company of Atting, Germany, decided that this 420 horsepower Lancair IV-P needed not three, not four, but <u>five</u> blades to efficiently transmit the horsepower and multiplied torque that the small-block Rodeck V-8 produces. The result is that the airplane will do 380 mph at 29,000 ft. (Editor's note: "WOW!")

DEFINITION
"PRO-PEL-LER,
also **PRO-PEL-LOR**,
1. one that propels
2. a device that consists of a central hub with radiating blades placed and twisted so that each forms part of a helical surface and that is used to propel a vehicle (such as a ship or an airplane)."

The definition given in the paragraph above pretty well defines all propellers, from the huge ones that move the ocean liner <u>Queen Elizabeth 2</u> to the small ones used on ultralight airplanes and even model airplanes. They come in many sizes and shapes, but the purpose is the same. To move an object through a non-solid medium, either water or air.

In fact, early writers and engineers called propellers "airscrews" because they screwed or bored their way through the air.

TOUGH DECISIONS

In every new aircraft design and new airframe power plant design, the propeller is the least definite thing about the new design. Without exception, every experimenter that ever flew a new design found that the first, second, and often the third choices of propellers was not the final choice. In many cases, the evolution of prop selection was based on flight tests and not ground testing. What works on a test stand on the ground often does not work when the airplane is flying through air at 100 miles per hour and faster. This chapter will help you narrow down your choices.

DESIGN THOUGHTS

This chapter does not mean to infer that there is one right and perfect way to design a perfect propeller for your auto engine that operates through a torque multiplying PSRU drive unit, belt, chain or gears. But with the illustrations and pictures shown here you should be able to make an intelligent choice.

With the thought provoking drawings and facts presented in this chapter, some thoughtful, enterprising inventor will likely design a much better propeller in the next 5 to 10 years. One thing that is certain now, none of the present designs in piston engine props are 100% or even 95% efficient.

One experimenter is presently working on a 2 speed PSRU that could show some merit. His thoughts are that once the airplane climbs to cruising altitude, the engine could be shifted into overdrive to reduce the rpm's of the engine by 50%. With the right design of a controllable pitch and CONTROLLABLE AIRFOIL propeller blade, his prop design might work. Only testing will tell the real story.

To be sure, a propeller that bites the air molecules the most efficiently will be the propeller that will lift the airplane the best, and move it forward the fastest. This means that the blade airfoil design is the elusive answer.

The Thunder Mustang also uses a custom made MT prop to produce very impressive climb and top speed performance. This is NOT your typical Bonanza prop!

When you look at a 2-blade propeller on a factory built, certified airplane such as a Piper Archer or a Cessna 172, it seems almost miraculous that such a skinny little piece of metal could actually move those relatively large airplanes through the sky at 100 to 125 miles per hour! But it does, and we do have a baseline design for propeller choices for airplanes of similar sizes, speeds and horsepower.

FIRST PROP CHOICES

If you are selecting a propeller for an experimental airplane that has not flown in the past, you will have to make an educated guess as to the correct propeller to use. Throughout this book, every airplane that is pictured is flying successfully. They are not 100% efficient, but they fly acceptably. You can choose a prop using these examples.

The easiest way to choose a suitable prop is to call one of the propeller suppliers that is listed in Chapter 19, and ask for technical advice. They will want to know several things, such as the kind of airplane, the cruise speed expected, the horsepower of the auto engine, and the reduction drive ratio. They will also want to know what the required and desired diameter of the prop should be. From the information that you give the propeller manufacturer, they can recommend a good choice of a propeller for your airplane.

THREE BLADE, TWO BLADE

Almost all the experimenters and inventors who fly V-6 and V-8 auto engines of less than 250 horsepower, report that they experienced significant vibration when they were flying with a 2 blade prop, but when they changed to a 3 blade prop, the vibration went away.

And then there is the aesthetic value of a 3 blade prop. Three blade props just LOOK better.

Unless you are planning to fly a small, ultra-light airplane with a low horsepower motor, you should consider a 3 blade prop, or even more blades. Several airplanes pictured in this book have 4 blade, 5 blade, and even 6 blade props. We can't get into the dynamics of propeller design in this chapter, because there is just too much to discuss, related to propeller design. Perhaps someone will write a 160 page book about propeller design and explain all the details of what makes a propeller work.

WOOD PROPS

A very high percentage of prop failures have been reported by pilots flying auto engines, both with 2 blade and with 3 blade wood props. This even applies to 2 blade props on 4 cylinder 50 horsepower VW engines. Save time and money and go directly to a composite prop.

HOFFMAN PROPS

Another European manufacturer, Hoffman, makes custom designed props for almost any size and horsepower auto engine powered experimental airplane. These props are also used on certified airplanes.

MT PROPS

The German prop manufacturer has been building composite, electric or hydraulic props for several years. The beautiful 5 blade prop used on the small block Chevrolet derived V-8 Lancair IV-P on page 117 is a composite MT Prop. Several airplanes shown in this book fly with MT Props. The V-12 Thunder Mustang pictured in this chapter has a custom designed 4 blade MT prop and it performs very well.

Rumor has it that this three blade prop on "Rare Bear" is a modified prop from a P-3 Orion, but whatever it came from, it makes the Grumman Bearcat the fastest prop driven airplane in the free world at nearly 500 mph.

RARE BEAR PROP

Without a doubt, the most impressive propeller ever seen on an airplane in the free world, is the 3 blade prop on the race plane, RARE BEAR, a WW II Grumman Bearcat single engine fighter plane. RARE BEAR is the fastest airplane in the free world that flies with a single engine and conventional propeller.

There could be some design ideas for future auto engine propeller engineering in analyzing the propeller used on RARE BEAR. Notice that the blades are extremely wide, all the way from the tip to the hub. Also notice that the blade pitch changes drastically from the hub out to the tip of the blade. The slower moving hub has a lot steeper pitch than the top of the blade does. This blade design theory is proven in the drawing on page 121 that shows the speed of different parts of a propeller blade at 2,750 rpm. You

can't buy a prop like this today, but look for it soon.

RUSSIAN BEAR PROPS

There is a Russian built multi-engine bomber that is propeller driven, and that is reported to be almost super-sonic. It uses counter-rotating props turning on common crankshafts on each engine. The counter-rotating props are reported to make so much noise that jets flying 500 feet way from the Bear bomber can't stand the noise. Although these props are evidently very efficient, you could not put up with that noise in your experimental aircraft. Another exotic airplane that uses the counter-rotating props is the Bugatti Racer. This design "BEARS" more investigation!

PROP TIP SPEED EXAMPLES

4.5 ft. prop length (54")	=	14.13 ft. travel per revolution
5.0 ft. prop length (60")	=	15.7 ft. travel per revolution
5.5 ft. prop length (66")	=	17.27 ft. travel per revolution
6.0 ft. prop length (72")	=	18.84 ft. travel per revolution

To find the maximum R.P.M. you can turn each of these props of different lengths, divide the distance they travel into 66,984 ft. (the speed of sound) and the answer will be the maximum RPM.

Example for the 5.0 ft. (60") prop @ 15.7 ft. travel:

$$66,984 \div 15.7 = 4,266 \text{ RPM maximum}$$

Example for the 6.0 ft. (72") prop @ 18.84 ft. travel:

$$66,984 \div 18.84 = 3,555 \text{ RPM maximum}$$

SPEED OF SOUND

The speed of sound at sea level, on a standard day, 59.0° F (15° C) is:

661.5	knots
761.2	miles per hour
66,984.0	ft. per minute
1,116.4	ft. per second

RELATIVE SPEEDS

88 ft. per second =	60 MPH
5,280 ft. per minute =	60 MPH
1 mi. per minute =	60 MPH

PROPELLER DYNAMICS NOTES
To see what your propeller is doing in flight at 2,750 rpm and at 200 mph, consider the notes given here about a theoretical 7 ft.diameter propeller. Note that the prop tip speed is 687 miles per hour plus 200 miles per hour forward speed. Those are some pretty wild dynamics.

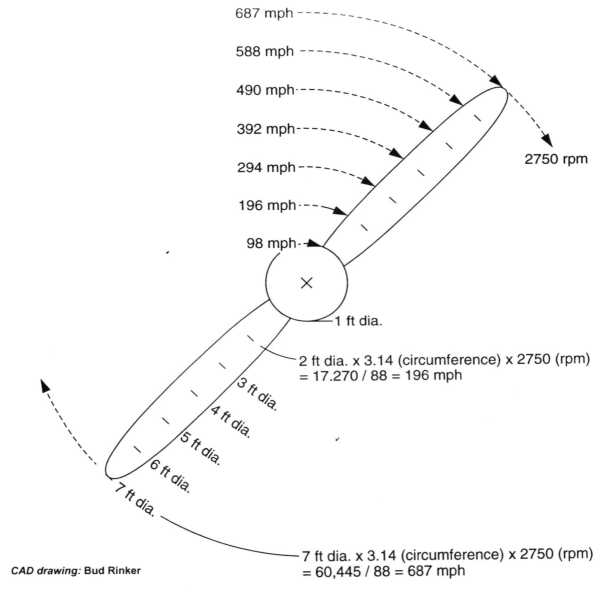

687 mph

588 mph

490 mph

392 mph

294 mph

196 mph

98 mph

2750 rpm

1 ft dia.

2 ft dia. x 3.14 (circumference) x 2750 (rpm)
= 17.270 / 88 = 196 mph

3 ft dia.
4 ft dia.
5 ft dia.
6 ft dia.
7 ft dia.

7 ft dia. x 3.14 (circumference) x 2750 (rpm)
= 60,445 / 88 = 687 mph

CAD drawing: Bud Rinker

761.2 mph = Speed of Sound (sea level)   (5280 ft per mile / 88 = 60 mph

At Prop Diameter:			Speed:
1 ft. x 3.14 x 2,750 =	8,635 divided by 88	=	98 mph
2 ft. x 6.28 x 2,750 =	17,270 divided by 88	=	196 mph
3 ft. x 9.42 x 2,750 =	25,905 divided by 88	=	294 mph
4 ft. x 12.56 x 2,750 =	34,540 divided by 88	=	392 mph
5 ft. x 15.7 x 2,750 =	43,175 divided by 88	=	490 mph
6 ft. x 18.84 x2,750 =	51,810 divided by 88	=	589 mph
7 ft. x 21.98 x2,750 =	60,445 divided by 88	=	687 mph

PROPELLER CHOICES

HELIX/PITCH...........THINK OF SCREW THREADS

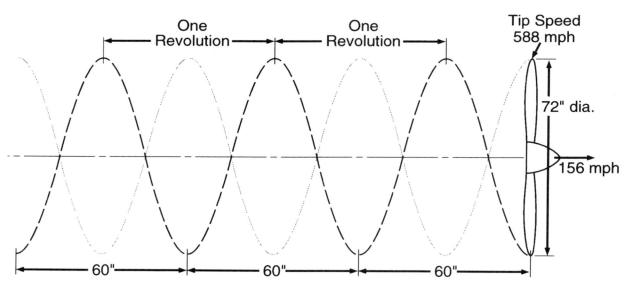

One Revolution — One Revolution

Tip Speed 588 mph

72" dia.

156 mph

60" — 60" — 60"

CAD drawing: Bud Rinker

PROPELLER RELIABILITY

As this book goes to press, there are comments in the movement about the dependability of experimental propellers. They are not perfect yet, but if you own a 250 horsepower brand "H" prop, you are also subjected to A.D.'s.

All 250 hp constant speed CERTIFIED brand "H" props must be removed every 1,000 hours or 5 years for probable blade cracking. The cost of this A.D. is $8,000, and as soon as that A.D. is complied with, another $3,500 A.D. hits you.

Today's experimental props are less than perfect, but they are better than the heavy, expensive, and failure prone certified props. Have faith, propeller technology WILL improve.

When a 60" pitch prop makes 1 rotation, it **COULD** advance 60". But it is not 100% efficient, so it does not move the full 60". In theory, a 60" pitch, <u>72" diameter propeller</u> turning <u>2750 rpm,</u> would propel an airplane 165,000 inches in one minute, or 13,750 feet in one minute, which equals<u>156 miles per hour.</u>
A 90" pitch prop, turning 2,750 rpm, advances 247,500 inches in one minute, = 20,625 ft. in one minute = 234 mph. But slip affects the prop. The propeller is not efficient.
One blade, 2,3,4,5, or even 6 blades set at 90" pitch still only advance just so much. The way to go faster is with more pitch (and more blade efficiency).

Both airplanes pictured on this page have electric in-flight adjustable IVO-Props, and the owners just love them. The prop on this Lancair ES is electrically actuated.

John Harlow, Jr. Photo

The obvious thing in this photo is how much skinnier the Cessna 172 fuselage is than the Lancair ES in the photo above. Elwyn Johnson has converted this prop to a hydraulic unit that is operated by the pilot, like a miniature hydraulic jack on the instrument panel.

Elwyn Johnson Photo

IVO PROPS

One manufacturer of experimental aircraft propellers is a company in the Los Angeles, California area, called IVO Prop. The name comes from its owner, Mr. Ivo Zdarsky, a refugee from communist Czechoslovakia, who escaped from his homeland in an ultra-lite trike powered by a 2 cycle engine and pushed by one of his prop designs. Ivo has developed a very unique pitch adjustment method for his props.

He clamps each blade with two bolts through the hub area, and runs a steel torsion rod about 2/3rds of the length of the blade, inside the center of the blade. Each end of the rod has a 90 degree lever attached. When the rod is twisted at the blade hub end, it causes the propeller blade to twist and change pitch at the 2/3rds part of the blade. Pitch changes can be made on the ground with 2 wrenches, or in flight by an electric motor and a planetary gear set built into the hub of the prop.

This prop design has been well proven in thousands of applications in ultra-light airplanes, and in hundreds of applications in auto engine powered air boats and auto engine powered snow vehicles. This chapter also pictures several experimental aircraft with auto engine power that are flying successfully with this unique design propeller.

One other advantage to this kind of prop is the relatively low weight. A three blade in-flight adjustable IvoProp weighs 27 pounds, which is about 5 pounds less than a 150 horsepower 2 blade fixed pitch aluminum prop. The 3 blade IvoProp is rated for up to 450 horsepower. Six blade, ground adjustable IvoProps are rated for up to 750 horsepower.

Another view of Elwyn Johnson's IVO-Prop that he operates with a miniature hydraulic jack on the Cessna 172 instrument panel. This engine is a 4.3 liter Chevrolet V-6 marine engine with a balance shaft and aluminum cylinder heads.

Here is a manual adjusting hub and one blade from an IVO-Prop. The small lever attached to a 4130 steel rod running through the prop blade, twists the prop and changes the pitch when the center screw is turned in or out.

IVO PROPS

It is important to discuss the fact that every manufacturer of experimental propellers has experienced blade failure. But it is also important to be fair and realize that every manufacturer of CERTIFIED propellers has ALSO experienced blade failures. and quite likely, if you search out the NTSB and FAA records, you will discover that the failure rates per 1,000 or 5,000 props is about the same for all prop manufacturers, both experimental and certified.

Things to inspect at every pre-flight of individual blade, ground adjustable or in-flight adjustable props, is the prop bolt torque. Each manufacturer will have their own pre-flight inspection procedure.

For the IVO-Prop, a piece of strain gauge stainless steel tape is placed across the gap between blades at the time the prop is installed and torqued down. At each pre-flight, the pilot must inspect the strain gauge to see if it shows signs of parting or cracking at the blade gap. If parting of the tape is evident, the prop bolts must be re-torqued to maximum torque before flight.

If the prop bolts get loose, the prop blades can work back and forth and cause the bolts to break. And of course, ALL, ALL, ALL experimental props, engines, and engine mounts should be secured by a formula one race plane type safety cable to prevent losing the engine if the prop breaks.

NEW AIRFOILS

Too late to include in this book, was a new air foil design prop displayed at the Oshkosh Fly-In this year. This new design prop blade is similar to two props, one slightly over laid over the other. Initial testing indicates a much quieter prop sound at full power operation. Look for good news from this new (patented) design.

SLIP RINGS FOR ELECTRIC PROPS

The two IvoProp installations shown on this page are electric, in-flight adjustable. The apparent range of in-flight pitch adjustment is 60 inches, from about 30 inches of pitch for take off to 90 inches of pitch for high speed cruise flight, a static pitch of 60 inches is built into the prop.

In order to effect a pitch change, an electric slip ring and carbon brushes is incorporated. Because every experimental engine installation is somewhat different, the builder must make his own slip ring brush bracket. The brush bracket must be adjustable by shims or by screw adjustments. The most important thing is that the brush bracket must be rigidly bound so that it will not vibrate while in flight. Any vibration would quickly wear out the two carbon brushes.

FIXED PITCH PROPS

You would think that a good prop selection for a reduction drive engine of a specific horsepower would be a fixed pitch prop off a certified direct drive aircraft engine of the same horsepower. But that is not a workable solution. The aircraft engine matched prop will always over speed and cavitate when bolted onto a reduction drive engine of the same horsepower. The reason for this is that the speed reduction unit multiplies torque over the torque produced by a direct drive certified aircraft engine.

Here is my electric IVO-Prop brush bracket and brush slip ring. The two aluminum plates transmit plus or minus (positive or negative) 12 volt current to the prop motor to change the blade pitch in flight. This is on my Buick V-6.

Timothy England uses a larger brush holder bracket to support the brush block for his electric in-flight adjustable prop. His engine is a 4.3 Chevrolet V-6. and his PSRU is an 8 mm cog belt drive by Northwest Aero Products.

IVO-Prop can sell you this 6 blade Magnum prop for your supercharged 502 cubic inch Chevrolet V-8 on your air boat. Each blade will adsorb 150 horsepower, making it capable of handling 900 horsepower. But Ivo de-rates this prop to 700 horsepower.

PROP CAVITATION

A number of auto engi converters report prop cavitati with small, thin blade props, ev with 3,4,5, and 6 blade version This characteristic exhibits itself the beginning of the take-off roll a at the high rpm operation range cruise.

At take-off, the prop starts pull, then almost quits pulling t airplane forward, and starts maki a loud "WOPPING" noise. If y have any experience w motorboats, you can relate this the prop cavitating the water a almost refusing to move. The san characteristic is evident at crui speed in an airplane when t throttle is advanced to go faster th the prop can handle.

The solution to th annoying trait is to change to a pr with wider blades. Again, take look at the very wide blades RARE BEAR in this chapter.

WARP-DRIVE PROPS

Several airplanes pictured this book are flying successfully w 3 to 6 blade Warp-Drive props. T RV-6A Chevrolet V-6 pictured on t front cover is flying with a 3 blac Warp-Drive prop. The Adventur Amphibian Chevrolet V-8 picture on page 128 flies quite well with a blade Warp-Drive prop. Presentl most of the airplanes pictured in th book, with Warp-Drive props, a ground adjustable units.

Warp-Drive rates their prop at about 50 to 70 horsepower p blade. A typical 2 blade warp-driv prop will fly nicely with 100 to 11 horsepower.

On page 128, a new, i flight adjustable Warp-Drive electr prop hub is shown. Several of thes new props should be flying soon.

Reiner Hoffman, of Stratus, Inc., flew this Subaru EA-81 powered Cessna 150 airplane for several years with the 3 blade, ground adjustable, Warp Drive prop. Performance figures indicated 115 to 125 horsepower.

PROP STALLING

Boat people call this situation "cavitation", see page 126. Airplane people call it stalling the prop. Either way, the prop is not doing its job when this situation happens.

The stalling or cavitation is controllable if you have an inflight adjustable prop. You simply flatten the blade pitch to overcome the stall. If you relate this to a wing stall, it would be like an accelerated stall. As you licensed pilots know, an accelerated stall occurs at a higher speed than a gentle stall. And therefore, prop stalling occurs at higher prop speeds, high angles of attack.

THEY ARE DIFFERENT

Any engine that transmits power through a reduction drive multiplies torque at its output. My high school physics teacher taught us that when you add a gear reduction to a motor or an engine, you can multiply the torque output of the engine, but you cannot multiply the horsepower. Therefore, if you add a 1.63 to 1 reduction drive unit to a 200 horsepower engine that produces 250 foot pounds of torque and 200 horsepower at 4,800 rpm's, you theoretically multiply the torque by 1.63 to 408 ft. pounds, but the horsepower stays the same.

THE REAL WORLD

What this means in the real world of propeller selection is that a reduction drive equipped engine has the extra torque to swing a larger diameter prop than a direct drive engine of equal horsepower. Many years and many hours of experimenting by many people have proven this real-world fact. But, large diameter props will exceed the sonic limits (the speed of sound) at the tips, so, the present solution is to add propeller blades to help adsorb the extra torque of the reduction drive engine.

This Adventure Amphibian is powered by one of the 350 Chevrolet V-8 "crate motors" and is pushed by this six-blade Warp Drive ground adjustable prop. In-flight adjustment provisions would improve the performance.

Warp-Drive Propeller Company displays this cut-away working model of their new electric in-flight adjustable propeller.

This picture shows the back side of the prop electric slip ring and control stick toggle switch operation of the Warp-Drive Electric in-flight adjustable prop.

This cut-away of a 3 blade certified hydraulic prop hub shows the possibilities for making a new composite blade, electric hub design.

A new design, die-cast, carbon fiber, scimitar type prop promises to change the pitch in flight by centrifugal force blade warping. Shades of past inventions!

The back side of the Jeff Ackland Legend hydraulic prop shows how the modified Cessna Conquest prop was adapted to automotive power plant use.

Engine Cowling Designs

◆◆◆◆◆◆◆◆◆◆◆◆◆◆◆◆◆◆◆◆◆◆◆◆◆◆◆◆◆◆◆◆◆◆◆◆

George Morse, Prowler Aviation, Redding, California, fabricated this very sleek aluminum cowl to take full advantage of the compact size of his Rodeck V-8, small block engine. The cowling is made in several pieces, screwed and riveted together. Aluminum cowls are easier and faster to build than fiberglass cowls. That is, if you can weld and form aluminum.

ALUMINUM COWLINGS

If fiberglass is just not your thing, you can do a pretty good job with aluminum if you have the patience and a little bit of "tin-bending" skills. You can use 2024-T on the sides of the cowl if you do not have any compound curves in that area.

For the most forward part of the cowl, where you want pretty bulges and compound curves, use 6061-0 (soft) aluminum about .032 inch to .040 inch thick. You can do a lot of stretching, shrinking and forming of 6061-0, and if you decide to weld it, 6061 is weldable, whereas 2024-T4 is not. After you shape the pretty nose pieces, you can then rivet the 6061-0 to the 2024-T.

Dr. Bill Harrold, Tom Jones (shown here), and Jess Meyers have made this Chevy V-6 conversion to their RV-6A airplane, a real work of art, and a highly efficient installation. No changes were needed to the top of the cowling.

ADAPTING AN EXISTING COWLING

The easiest solution to the cowling problem is to adapt an existing cowling to house the auto engine. By using splices, making cut-outs, and blanking off unneeded holes, a suitable cowling can be adapted to fit.

Let's say you plan to install a Ford Fiesta or Escort engine in the neat little Zenith 2-place, low-wing airplane. Zenair makes a fiberglass cowl for that airplane that has more than enough room for the Ford engine. All you would need to do is open up parts of the fiberglass cowl for the Ford air cleaner to stick up through, and then build a fiberglass bubble to enclose the part that extends through the cowl.

I have also adapted cowls from different airplanes by sawing them lengthwise into four sections, like an orange is peeled, and then splice the four pieces back together, either smaller or larger as needed for the particular airplane I am working on. Even if you saw out holes and slots in 50% of the cowling, you still have a basic form to work with so you do not have to start from scratch.

The Jeff Ackland Legend airplane required a fully fabricated fiberglass cowling to effectively follow the fuselage lines.

The only change required to the cowl on Jess Meyer's RV-6A was to add a small bulge under the propeller to cover the full-size Chevrolet V-6 flywheel flex plate. A smaller flex plate from a NISSAN Sentra would even eliminate this one change to the cowling.

There is a Blanton conversion Ford 3.8 V-6 engine under the cowl of this Glassair T.D. airplane. The small bulge under the prop spinner was the only change made to the cowling. Exhaust pipe cut-outs are visible under the cowl.

After I installed the Buick V-6 engine in my 1972 Grumann Traveler 4 place airplane, I stretched strips of masking tape from the firewall to the prop spinner to see how well the new nose cowling would fit around the engine.

FABRICATE YOUR OWN COWLING FROM FOAM

Take a good look at the various pictures of my Buick V-6 engine cowl design, pictured in this chapter. If you want a custom, slick cowl, this is the best way to do it:

1. Cover your engine with plastic garbage bags to keep the foam dust out while you are filing and sanding.
2. By using cardboard boxes, masking tape and other "filable" materials, construct a rough shape of your cowl in place on your airplane.
3. Cover the cardboard, etc., with blocks of urethane or Styrofoam like they use to build Variezes. Glue the foam blocks together.
4. Sand and form the cowl foam to the shape you want. Be sure to include air inlets and outlet holes. Make the foam very smooth.
5. Cover the foam cowl with several layers of fiberglass cloth.
6. After the fiberglass has dried, you can use a small circular saw blade in a 1/4 inch drill motor to split the cowl so you can remove it from the engine.
7. Fit the cowl with Dzus fasteners so you can check the oil and water easily, and so you can remove it from the engine.

This will provide the nicest-looking cowl, but expect to spend 3 or 4 weeks building it. A full cowling could take four to six weeks, depending on the amount of time you are able to devote to this part of your project.

The thing that I dislike most about fiberglass cowls is all the dust that you generate by sanding them to make them smooth. An aluminum cowl is already smooth when you start building it, and even if you weld and rivet it, you will stay a lot cleaner than you would if you make a fiberglass cowl.

At this point in my new cowl construction, sheets of aluminum would have been easier to form than these foam sheets. You can use large sheets of corrugated cardboard to make the patterns for aluminum sheets.

ATTACHING THE COWL TO THE FUSELAGE

Take a look at a factory-built airplane cowl attachment. Usually there will be about 12 to 15 sheet metal screws holding the cowl in place, or in some cases there will be about 8 to 10 camlock or Dzus fasteners. If you attach your cowl similar to an **EQUIVALENT** speed factory airplane, it will be sufficient, I personally like to use camlock fasteners because it is easier to remove and replace the cowl if you don't have screws to get lost.

Check out the pictures of auto engine cowls in this book and pick one that suits your airplane and engine installation, and use that cowl as a guideline. Don't forget to allow a reasonable amount of airflow around the engine. Pay close attention to **HEAT** in the engine compartment and make sure the distributor, carburetor, alternator, fuel lines and any oil lines have sufficient air flow to keep them from overheating. You may want to run a special duct just to cool the alternator and fuel system parts. Also, allow for air to get out of the engine compartment after shut down so the engine won't heat soak while sitting still.

Don't forget about the effects of flying through rain. If you have vents directed at your ignition system, you could have water-short problems in the rain. You will need a water deflector to protect ignition and electronic parts from rain. Water deflectors can be made from small pieces of aluminum, clamped securely to prevent them from vibrating and breaking off in flight.

Flight tests for the first 40 to 50 hours are supposed to give you time to check for problems before you head out with the family on a long, cross country trip. Use these hours of restriction to thoroughly check everything out.

For more information on how to build auto engine nose cowls, refer to Tony Bingeli's book, *The Sport Plane Builder,* available from E.A.A. Publications, Oshkosh, WI.

At this point, I am ready to sand the corners smooth and apply fiberglass. Also, at this point, sheets of corrugated cardboard would have been just as good as the foam. And at this point, sheets of 6061-O x .050" aluminum would have faired into a good shape just as well as this foam did.

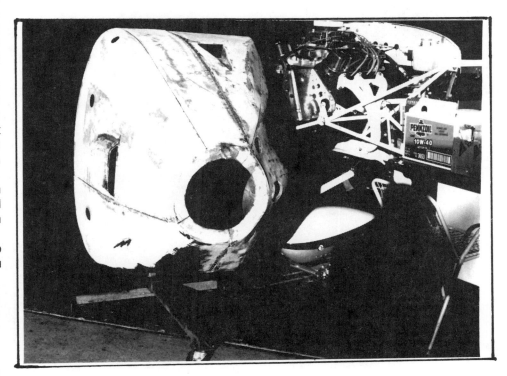

The almost finished nose cowl is mounted to a plywood sheet the exact size of the firewall, and the assembly is mounted in an auto engine work stand so I can rotate it while sanding and Bondo-ing to make it smooth. The top of the cowl is to your left in the picture. Welded aluminum would have been 10 times faster!

133

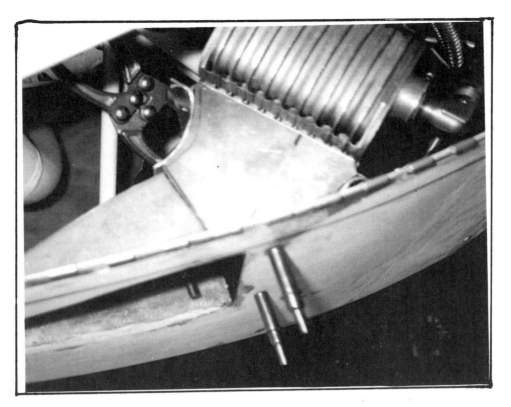

Here I am fabricating a NACA duct for the engine air cooler after the cowling is finished. I will also use the hot air off the oil cooler as a heater and defroster for the cockpit.

This Glassair T.D. has a Ford V-6 engine in the cowling. Three exhaust pipes come out each side. A P-51 type radiator is mounted under the belly. The nose cowl is almost, but not quite, stock.

ARE COWLINGS COMPLICATED?

Many people shy away from building the cowling for an experimental airplane. They either buy a prefabricated cowl or they send the work out to a specialist. The big problem with buying a prefabricated cowling is that most prefab cowls are made to fit an air-cooled airplane engine, and they will be much wider than necessary for the narrower auto engine.

The big problem with sending the work out to a cowling specialist is that the cowling could cost almost as much as the entire auto engine!

Another problem with cowlings is that they usually need to be aerodynamically smooth because they "break the wind" for the entire fuselage in a single engine airplane with the engine in the nose. In other words, cowlings need to be streamlined and the contours of the cowling must be compound curves, not single curves.

Take heart! Cowlings are not as hard to make as it seems at first. I have several solutions to the cowling dilemma. I'll tell you how.

TYPES OF COWLS

For custom made cowls, fiberglass or sheet aluminum works good. Sheet aluminum, 6062-0 & 6061-T6 x .032" thick will work well. This all depends on the airplane that you are making the cowling for. If you want to make a really nice cowl, you may need to buy a video tape about forming aluminum with an English Wheel.

In other cases, where you already have a fiberglass nose cowl off an RV-6 or Glassair or Lancair for instance, it is much faster to just modify your existing cowl. In any case, remember that your auto engine is a lot more narrow than an air cooled opposed aircraft engine. You will always have more cowl space with an auto than with an aircraft engine.

Ground Testing Auto Engines

◆◆

You are looking at some very serious horsepower here. Falconer Racing Engines, of Salinas, California, is dyno testing one of their new V-12 small block engines for Dan Denny's P-51 Mustang replica. Dyno testing is a very good idea if you want to know your engine's power output before you fly.

GROUND TEST FIRST

If you have access to an engine stand dynamometer in your area, by all means check the cost of 4 hours test time on a dynamometer. With the electronic computerized dyno sheet print outs available today, you will be able to tell very quickly what the horsepower of your engine is, at all rpm's, and this information will help you decide on the best prop for your airplane, the best timing for the ignition, and the best adjustments for your fuel injection or your carburetor.

In years past, many engines were burned up or otherwise ruined by inexperienced people running them on dynamometers. Be sure that your choice of dyno shops 135

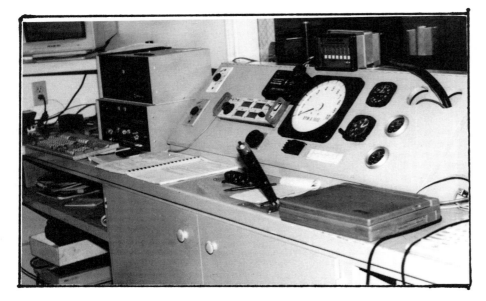

floor, and go out into the country to ground test your engine. This chapter shows 5 different auto and motorcycle engines set up on trailer mounted test stands. Copy these designs to make your trailer test stand work well.

Once you mount your engine on a trailer, you can also become a bona fide exhibitor at air shows and fly-ins. But make sure that no pets, no friends, and no strangers can get into the propeller when the engine is running.

The dynamometer control room at Duttweiler's place in Saticoy, California, is completely electronic. Computers control the dyno runs and computers print out complete results during and after the runs.

includes a dyno operator that is experienced and that is not prone to breaking engines.

DYNO TEST PSRU'S

It is relatively easy to fabricate an adapter that will connect the propeller hub of your PSRU to the dynamometer flex plate adapter. Most engine dynos readily connect the dyno to the flywheel or flex plate of the engine to be tested. The main problem of testing your engine with the PSRU attached, is the usual offset of 8" to 10" of the prop shaft to the engine crankshaft. You will have to mount your engine 8" to 10" lower than you would if you test through the crankshaft only.

TRAILER TESTING

Auto wrecking yards (the new ones call themselves RECYCLERS), usually start and run salvage engines in the dirt or on the floor, without a radiator to show the potential buyer that the engine will indeed run. It only takes 15 minutes or so to jump a battery to the engine and add gasoline to the fuel line.

Reasoning that the Recyclers can run an engine with this little amount of trouble, you should be able to hook up a radiator, mount the engine to a utility trailer

Al Harralson built this trailer mounted test stand and control panel (in lower photo) to test his single rotor Mazda engine and propeller.

This water cooled 4 cylinder Honda Gold Wing Motorcycle on a test stand proved three things: *1. The engine was 25% heavier than a Honda car engine of the same c.c.'s*
2. The engine was extremely loud when running.
3. The horsepower was much lower than expected.
Moral: Ground test before you fly.

AIRPLANE TEST STAND RUNNING

Making the very wise assumption that any and all auto engines will run and produce power reasonably close to the advertised horsepower, you can also complete your auto engine conversion, mount it in your airplane, and do your ground testing, using the airplane as a ground test fixture. In fact, you will enjoy making taxi tests, high speed taxi tests, and extended duration ground runs while you really "feel out" the engine actually mounted in your airplane!

TWO PRIORITIES

So that you don't wreck your dream engine, make sure that you have the two most important systems worked out very well. The two prime systems that can make or break your conversion are:
1. **cooling**, and
2. **oiling**.
Always monitor coolant temperature and oil pressure and temperature.

FIRE DANGER

Never forget that gasoline is extremely flammable, and that fire is a constant possibility when testing new engine systems. Always keep an A.B.C. rated fire extinguisher ready for use in case a fuel leak, oil leak, or even alcohol antifreeze leak should cause a fire.

DON'T WEAR IT OUT

You only need to ground test your engine just enough to work out the bugs. Once you have it starting and running right, it is time to stop wearing it out on the test stand and go fly it!;

Mike Middleton, of Boise, Idaho, tested this Geo Metro engine on this test stand and was quite pleased with the results. Another builder later flew a P-38 replica with two of these engines for power plants, and they performed very well.

BENEFITS OF GROUND TESTING

It is relatively easy and fast to set up almost any liquid cooled engine on a portable test stand. You should have adjustable engine mount adapters, and you should also have several spare sizes of radiator hose adapters.

The radiator should be equipped with an electric fan for proper cooling, and the test stand should have a sturdy battery tray. You could even borrow the battery out of your car for ground testing.

And be sensible about your fuel supply. A good choice for a fuel tank would be a 20 gallon race car fuel cell that would actually cost less than an un-safe jerry can. And be sure to include a fire extinguisher!

BUILDING YOUR TEST STAND

Having test stand run a large number of engines I can suggest to you that you make your test stand adaptable to both a trailer and to a pickup truck bed. A very durable engine test stand can be made from square tubing, 1 1/2" outside dimension with 1/8" (.125") wall thickness. You can make a temporary engine stand from 1" O.D. E.M.T. electrical tubing, but it may not stand up to heavy engines and more than 150 horsepower. The square tubing frame is best.

You will find that your neighbors will not care for long hours of test stand running. They may smile and give you a thumbs up when you first start it up, but after 10 minutes of 100% power runs and all that prop noise, they will get tired of it. You will want to be able to tow or haul your engine out to a remote area where noise is less of a problem.

Remember that propeller noise can carry for several miles. Even at an airport, your fellow pilots will not like to hear 40, 60, or 80 hours of prop noise. In my own situation, I can go to a remote mountain top road about 20 miles from my house and shop to do engines runs. But if I get noise complaints, I must be a good neighbor and move on. A trailer or a pickup truck is the only way to do this.

FIRE EXTINGUISHING

Don't even start to install an engine on a test stand until you have made sure that a fire extinguisher is available. A couple of 2 1/2 pound dry chemical extinguisher will sure be a lot better in case of fire, than trying to throw sand and gravel at a fire. Cleaning up dry chemical is a pain, but is far better than letting your pretty new engine burn!

This trailer mounted test stand allowed the propeller designer to take the engine to the airport for testing, and to rack up dozens of hours of full power running to test the dependability of the electric propeller.

There can't be very much expense involved in converting this 95 horsepower Corvair engine to direct drive, carving out a wood test club prop, building a simple angle iron test stand and mounting everything on this trailer. If you wonder if a particular engine would be good to convert to aircraft use, do a test stand engine like this to find out for sure.

Flight Testing Auto Engines

At first glance, it appears that Jess Meyers is flying his Chevy V-6/RV-6A over some hostile terrain, but in reality, the Santa Paula, California, non-controlled airport is within gliding distance just off his left wing. When flight testing, never get further from a suitable airport or a grass strip, than your airplane will safely glide to WHEN, not if, you have an emergency. And the same is true for aircraft powered by CERTIFIED aircraft engines.

FLIGHT TESTING AUTO ENGINES

THE FIRST RUN

The first time you start and run your auto engine conversion in your airplane should be a very exciting occasion. Each start-up of a new airplane, even in an aircraft factory, is an exciting event, and the first engine run of your own engine will be very memorable to you. I will tell you how the factories prepare and go through their first runs of new engines. The procedure is the same whether the airplane is a single engine 100 hp model, a $350,000 twin, or a 5 million dollar turboprop airliner.

THE PROCEDURE
1. The airplane is parked in a safe place where prop-blast will not blow people or property away. The wheels should be chocked securely, and if necessary, tie the tail tie-downs to a tree, a post, or something substantial. Don't trust the brakes unless you want to take chances on un-planned take-offs!
2. The person preparing for the first run makes sure that fire trucks can get access to the airplane in case of a major fire during engine run.

139

It looks like Jess Meyers is flying over dangerous mountain terrain in this picture, but, in reality, he is 2,000 feet directly over the Santa Paula, California, non-controlled airport.

3. At least three (3) qualified people will be on duty to help and provide for safety:

A. The "crew chief" will sit in the pilot's seat to operate the throttle, watch the various gauges, oil pressure, oil temperature, cylinder head temperature, and be ready to shut everything down in case of oil leaks, fuel leaks, fire, or unusual noises. It is also a good idea to have an extra person in the airplane to watch for problems, and to write down the readings taken from the instrument panel. It really is important to write down the things that occur during that first engine run. You will find that things like erratic tachometer readings, low oil pressure, high coolant

A rigidly mounted video camera can fully record your instrument panel in your first test flights. Mount the camera to a roll bar or to the co-pilot's seat back and aim it for shots like this. Here we see: 140 knots, 400 ft. min. rate of climb, 1,340 ft. MSL, 3,200 engine rpm, 220°F. engine temp. 200°F. radiator temp. and 126.75 dialed in on the comm radio of this RV-6A Chevy V-6.

temperatures WILL occur, and it is wise to write these events down on paper so you can work on eliminating the problems before the next engine run.

B. Outside the airplane, about 20 feet in front of the engine and in plain, full view of the crew chief must be a "fire watch" person. He is always stationed there with a portable fire extinguisher in hand, ready to use if a fire should start. This person should never move from his "fire watch" position until the engine is shut down and the propeller stops turning. If a fire does break out during the first engine run, the "fire watch" should be ready to give the crew chief a "cut the engine" sign, and after the prop stops turning, be in a position to spray the fire with the portable fire extinguisher. If the fire is too big to put out with the portable extinguisher, the only thing left to do is to call the nearest fire department, and let them stop the fire from spreading to surrounding structures.

C. The third (or fourth) person should be in position, usually under the wing, to look for oil leaks and fuel leaks while the engine is running. Of course, the engine cowl is left off for the first engine run so you can detect ;leaks. etc. more easily. The ground observer should look for liquid such as oil, gasoline or water dripping off the engine during first run, and signal the "fire watch" to cut the engine as soon as any sign of leaks is found. The "fire watch", in turn, will signal the pilot to stop the engine. This "cut" signal is waving your hand from side to side at your throat as to indicate cutting your throat.

You can always expect to see excessive smoke coming out the exhaust tailpipe as soon as you start up any newly installed engine. This smoke normally lasts 30 to 60 seconds, but if it is still smoking heavily after two minutes, something is wrong. Stop the engine and find out why. You can also expect some smoke to come from the outside of the engine, especially in the area

Again, it looks like Reiner Hoffman in his Subaru powered Cessna 150 is flying over dangerous, but beautiful, terrain. Below his is beautiful Lake Cachuma, and at his 10:00 o'clock position is the Santa Ynez, California, non-controlled airport, again, within safe gliding distance.

near the exhaust ports on the cylinder heads. External smoke is caused by oil being on the engine because you have touched it with your oily hands, if you washed the engine with solvent before you started converting it for aircraft use, there will be some residual oil that will soon evaporate, and the smoking will soon stop.

CAUTION

Remember that turning propellers are _very_ dangerous. Never get close enough to a turning propeller that you could fall into it. Also, be very careful that your arms and hands do not come close enough to the turning propeller to accidentally get tangled up in it. Because of the possibility of a prop blade breaking and slinging off, never let anyone stand in the arc (to the side of) the prop while it is turning. Stay at least 200 feet away from the area where a broken prop blade could go.

In starting any newly-installed engine, whether it is a motorcycle, automobile, or an airplane engine in its "proper" place, oil and fuel leaks can occur. I always check for oil leaks every time I change the oil and oil filter in my family cars, so you should check for oil leaks in your freshly installed auto engine in your airplane. If you find oil or fuel leaks, stop the engine, and repair the leak immediately. Then restart the engine to verify the leak has been stopped.

DON'T FEEL BADLY IF YOUR ENGINE LEAKS

I have seen many gallons of oil and fuel spilled and leaked from *expensive airplane engines* on first engine runs, especially at the factories! It can happen to anyone.

GROUND RUNNING TIME

The first ground run should last from 5 minutes to as long as 15 minutes, but no longer. You should shut the engine off, and thoroughly inspect everything related to the installation to verify no problems. Look for leaks, loose bolts, and especially check the prop bolts for security. After a thorough check of everything, which should last at least 30 minutes, you can start up the engine again.

SECOND ENGINE RUN

The second time you run the engine, it is not necessary to have a "fire watch. and a "leak watch". It is a good idea to have someone in view of the crew chief so signals may be given if something goes wrong, but if the first engine run was relatively trouble-free, the second and later engine runs should be trouble-free also. The cowl should be on for the second and subsequent ground runs.

SUBSEQUENT GROUND RUNS

The airplane factories seldom run a new airplane engine more than 30 to 45 minutes on the ground before the first flight. Actually, the first 15 minutes of ground running should tell you all you need to know about how well things are going to stay together Any more than 45 minutes of ground running is just wearing out the engine and could cause overheating. After the first 3 to 5 minutes of run-time, you can run the engine at 3/4 to full rpm. The first 3 to 5 minutes should be run at 1/3 rpm, such as a maximum of 2000 rpm for an engine that would red-line at 6,000 rpm. Therefore, 1500 rpm would be a very good speed to run any auto engine for its first run.

ABORTED TAKE-OFFS

Before attempting the first flight, especially with a brand-new engine installation, at least five aborted take-offs should be performed. The runway should be at least three times longer than is required for lift off, since you will be getting the airplane to flying speed and then stopping as if you were landing. The purpose of aborted take-offs is to feel out the airplane to see if everything feels airworthy. Airplane engines run faster just as the airplane leaves the ground than they do during static run-ups. The procedure for doing an aborted take-off is the same as for a short field take-off. Start the take-off at the extreme end of the runway, and hold the brakes while advancing the throttle to full power. When full power is obtained, and the engine is running smoothly with no roughness, release the brakes and let the airplane accelerate to the speed where the main gear is ready to leave the runway. The instant the plane breaks ground, slowly but firmly retard the throttle to idle, and bring the airplane to a smooth stop, straight ahead. Don't take chances on running off the end of the runway, or you might ruin all that hard work you did in converting the engine to aircraft power.

A major advantage of doing at least five aborted take-offs is that you will get a very good idea of how well your engine's cowling is going to do aerodynamically. If the cowling flutters and flops around during the aborted take-off, you surely need to redesign it before you fly the airplane.

THE FIRST FLIGHT

Before this event occurs, you must have a signed copy of an airworthiness certificate in your hand, and the airworthiness certificate must be in the airplane during all flights. You can also expect a large crowd of friends and interested people to be on hand to witness the event.

Several years ago in Santa Maria, California, the local newspaper, radio, and the television stations were notified that a pilot was planning to make the first flight in a Fairchild High Wing with a Mazda RX-7 car engine installed. Needless to say, there was a very large crowd on hand to witness the event.

The pilot, engineer, mechanic, all-in-one, provided a very long picnic table full of food for all the people that came to watch the first flight of his Mazda rotary-powered Fairchild airplane. The plane flew, with a couple of problems, but safely, and the most excited person there was the pilot, of course. Everyone had a good time.

The two problems I mentioned were much less horsepower than expected, and very high oil temperatures. The engine had been race-car modified and was supposed to produce 275 horsepower, but the plane flew more like it had 100 hp, and because of this problem, the oil temperature rose to the number of the horsepower hoped for, 275° F!

OVER-MODIFICATION of this Mazda rotary engine caused reduced performance. The prop speed was reduced by a beautiful gearbox drive, and the custom fiberglass cowl was a beautiful thing to see. I am sorry to say that I left my camera home that day, so no pictures are available.

YOUR FLIGHT TEST

Quite likely, you will have to fly your auto engine in a 25-mile radius of your selected airport until you have accumulated 40 to 60 hours of flying time that is free of major trouble. You can do **minor** repairs, such as carburetor and timing adjustments, and even cowling vent changes, but if you blow the engine because all the water boiled away, you will have to fix the cause of the overheating, and start your 40-hour flight test all over again.

LONG RUNWAY

Many pilots of first flight experimental airplanes prefer to fly the airplane out of airports with long runways, in excess of 5,000 feet long, if possible. This gives them the chance to land straight ahead if anything should go wrong on take-off. However, if you allow 60% more stopping space than lift-off space, you could have someone stand at the 40% runway length point and abort the take-off at that point, no matter how short the runway is.

BE SCIENTIFIC

You should keep a clipboard and pencil in the airplane, and take instrument readings every flight to note trends in performance and to evaluate the results of any adjustments you make.

This system will make the solo time pass faster, and the numbers you record will be very valuable to you in evaluating your engine installation. *ENJOY YOUR AUTO ENGINE AIRPLANE!*

HOW TO "SELL" AN AUTO ENGINE TO YOUR GOVERNMENT AGENCY

Convincing the government representative who is authorized to sign off your airplane for flight, that an automobile engine is safe to fly, can be a difficult job, or it can be as easy as a phone call to your mother.

When I first told my FAA GADO representative from the Van Nuys, California office that I planned to install a Corvair automobile engine in my Cessna 150 airplane, his comment was, "Let me know what day you plane your first flight, and I will meet you at the airport with an airworthiness certificate." He told me he didn't even need to see the installation during the development time. He only wanted to inspect for safety wire, the absence of fuel leaks, and other things a mechanic would look for on a Lycoming or Continental airplane engine installation. Part of the reason that this particular FAA representative was so easy to convince of the safety of my automobile engine in an airplane was that I worked with him at Aerostar when I was a Research and Development Project Engineer. He knew I would have had a positive attitude toward *any* auto engine installation that showed basic safety in the design of the installation.

On the other hand, I have talked to FAA GADO representatives from other regions in the USA who would not under any circumstances issue an airworthiness certificate to an auto engine in an airplane.

So, getting approval to fly an auto engine is like anything else in this world. If at first you don't succeed, try a different government representative. In my work with "the authorities" as the various government agencies call themselves, I have found that each "authority" has his own personal opinion of what should be flown and what should not be flown. Generally speaking, I have found that Australia is the most difficult country to get homebuilt aircraft approval, but even so, the difficulty there is only based on safety considerations. If you can convince the Australian authorities that your auto engine-powered airplane is going to fly as dependably as any other engine, then obtaining an airworthiness approval should be possible.

If you have particular difficulty in convincing your local authority that your auto engine-powered airplane is safe to fly, it might help if you donate to his office a copy of this book. Ordinarily, if they have factual, written, back-up material in favor of a specific airworthiness item, they will be much more likely to be favorable toward the "experiment". In every case, they only want to avoid airplane accidents. Also remember that each experimental airplane must pass or fail on its own merits. How safely **you** make the engine installation is of prime importance, regardless of whether it is an auto engine or an airplane engine.

Another good contact is through your local Experimental Aircraft Association chapter. The chapter officers should be able to furnish the proper government contacts to help you obtain approval to fly your auto-powered airplane. If you are unable to find local EAA people to assist you, you can write to the EAA International Headquarters for assistance:

Experimental Aircraft Association
Wittman Field
Oshkosh, WI 54903-2591
USA

CAN YOU PUT AN AUTO ENGINE IN A CESSNA, PIPER, OR OTHER PRODUCTION AIRPLANE?

Yes. You can install an auto engine in your factory-built airplane, flight test it for about 100 hours, **ENDURE** about six months of submitting test data to the authorities, let two or three government representatives fly the airplane while trying to find fault with the engine, and if you don't make them mad at you, you can get a Supplemental Type Certificate to install that specific auto engine in that specific airplane make and model. In other words, you can get a license to install a 231 cubic inch Buick V-6 engine in a Cessna 172 airframe. You can install this engine in 1 or 1,000 or more airplanes of the same make and model. You can sell the S.T.C. to other people who want to do exactly the same thing.

An S.T.C. for a specific auto engine installation in a factory airplane would be exactly like getting an S.T.C. to install a taildragger kit in a Cessna 150, or to install a 150 hp or 180 hp Lycoming in a Cessna 150. It can be done, and it really is not as hard to do as many people think. I have supervised S.T.C. work, and the hardest part was providing the FAA all the data they needed. The actual flight test time to obtain an S.T.C. could be as little as 5 hours or as much as 500 hours. Flight test time is established by the local FAA office. Ninety-five percent of the time you will be told "It can't be done," but don't give up. Go see a different GADO (General Aviation District Office), and keep trying until you find a government office with a more open-minded attitude.

CERTIFY AN AUTO ENGINE?

I'll tell you right now that it is totally possible to certify an auto engine for use in production aircraft for less than $25,000 total cost, including salaries of the people required for running the FAA certification tests! If you did not know the requirements, I'll tell you that it only takes 150 hours of test stand time to certify a brand new engine design. That's equal to driving your car, towing a camping trailer, across the USA and back two times, or about 10,000 miles total! That is ALL!!!

FAA REQUIREMENTS

I found a copy of the *Federal Register*, January 1990 edition, in the college library at the University of California, Santa Barbara. On pages 660, 661, and 662, "Section, 33,29, subpart C - Design and Construction; Reciprocating Aircraft Engines", it tells how to certify a new style engine. Anyone can go to a large library and check out a copy of the

Federal Register. You can double-check my statements.

PARAGRAPH 33.35, FUEL AND INDUCTION SYSTEM

This paragraph essentially allows all modern (1934 - 1991) auto engine carburetors and mechanical and electronic fuel injection systems to be used.

PARAGRAPH 33.37, IGNITION SYSTEM

This paragraph tells us that an airplane engine must have dual ignition OR have an ignition system of equivalent in-flight reliability. The OR in the paragraph gives you the right to have single spark plugs and single distributors if you can prove that your single ignition is as reliable as a 1925 era designed dual magneto system. By historical records of aircraft ignition failures and auto ignition failures, you can prove this.

But, it is wise to have a dual electronic ignition capability, but not dual spark plugs (all new V-6 Buick engines, for instance, have TRIPLE ignition coils.

PARAGRAPH 33.39, LUBRICATION SYSTEM

This paragraph essentially tells us that the engine must be designed to incorporate an oil cooler. You could simply allow air to flow across and around the engine oil pan, and this would constitute cooling the oil, but a separate oil cooling radiator is a good idea. Race car oil coolers cost about $100, and you must run a crankcase vent.

PARAGRAPH 33.43, VIBRATION TEST

The FAA wants us to verify ten million stress reversals, which simply means that they want to know that a piston and connecting rod can change direction of movements that many times. I calculate that to be 41.67 hours of operation or 2,292 miles of highway driving! I guess this means that they would allow your engine to throw a rod at 2,293 miles, and still pass the test!

PARAGRAPH 33.47, DETONATION TEST

This paragraph simply states that you should run the engine rich enough to prevent detonation. ALL car engines will pass this test easily.

PARAGRAPH 33.49, ENDURANCE TEST

Let's assume you are testing a 220-horsepower auto engine. In general, an engine must operate a total of 150 (one hundred fifty) hours, total time, to be certified. The detonation test, vibration tests, and any other test, can be part of the 150-hour test, or deducted from parts of the 150-hour test. Basically, there are seven separate tests that last 20 hours, and one of the seven tests lasts 30 hours. Here are the tests:

1. A 30-hour run @ 5 minutes of 220 hp (take-off), and 5 minutes @ 121 hp.

2. A 20-hour run @ 1 1/2 hours @ 200 hp, alternated with 1 1/2 hours @ 121 hp.

3. 20 hours at max cruise power, usually 75% and alternate with 1 1/2 hour periods of 121 hp. (And so forth.)

SUMMARY

If two men decided to certify a car engine, they could do it without outside help. The biggest problem would be in observing and controlling the 20-hour-at-a-time test stand runs. Each man could watch the engine for 1 1/2 hours, then go take a nap while the other man watched the test, except for the first 30-hour test. With modern electronic engine controls, the engine's tachometer could operate the engine at all required power settings, automatically.

Of course, the FAA would have to furnish people to observe and validate the test. One hundred fifty hours is less than 1 month of normal work weeks. Now, can you see that it can be done for a few thousand dollars, not millions? Can you see how the aviation industry continually misleads us, and avoids new, less expensive, and for more modern ideas?

DO YOU WANT TO GET A STC?

It is completely possible to get a S.T.C. (Supplemental Type Certificate) to put a Chevrolet 300 horsepower V-8 engine in a Piper Cherokee 180, a Cessna 172, a Grumman Tiger, and all other airplanes that presently use IO-360 Lycoming engines.

First, you need to develop the engine package to the point that it will be dependable for NO MORE THAN 100 to 150 hours! That is all they require of a certified IO-360 Aircraft engine.

Next, you need to install one complete engine package into each of the airplanes that you want an STC for. At this point, you may want to just get an STC for your 300 horsepower V-8 in a Piper Cherokee 180. Then, later, you could install an identical engine package in one each of the other airplane make and models that you want STC's for. A Mooney 231 would fly great with 300 horsepower.

Next, the FAA GADO will want you to fly a series of tests, including stalls, fly-overs to record decibels of noise, take-offs over a 50 ft. obstacle, and about 40 hours of total flight time. The FAA does not charge for these tests because they are already paid by us tax payers.

The total cost to get a STC should be less than $75,000. Then you could sell STC's to hundreds of other Piper Cherokee 180 owners at $1000 for the STC paperwork.

In Chapter 20, I predicted that in the next 10 years, we will see CERTIFIED auto engines in factory airplanes. There is no technical or legal reason why this is a problem. Just picture being able to call your General Motors Dealer to have him deliver a new 350 horsepower V-8 engine to your hanger for only $2,995.00! It will happen.

CAUTION: MANDATORY READING BEFORE YOU FLY YOUR AUTOMOTIVE ENGINE!!

EXPLOSION-PROOF FUEL SYSTEMS

Many airplane crashes and hard landings are survivable, except that the pilot and passengers are burned to death in post-crash fires that are the end result.

Almost all present designs of aircraft fuel tanks are extremely prone to splitting open and dumping fuel, which is then ignited by the inevitable sparks that result from the airplane hitting things as it crashes. Even if the airplane lands on soft dirt, the ruptured fuel tanks can spill fuel on to hot engine parts and explode. Many airplanes are almost totally intact after hard landings or slow speed crashes, but the airplane's occupants are consumed by fire.

This form of tragedy can be prevented. The materials and technology exists today that can save hundreds of lives, and yes, even airplanes, by preventing post-crash fires and explosions.

Since 1969, race cars have used this system, and racing fatalities are now almost nonexistent as a direct result of modern technology fire protection practices incorporated in race car designs of today.

The method of explosion prevention and fire protection for airplanes is to fill the fuel tanks with reticulated polyurethane foam, and to mount a HALON fire extinguishing system in the passenger compartment. The 2% density foam acts as a mechanical barrier to prevent explosions, and as a heat-sink to remove energy from any possible fire,

The cost of the polyurethane foam is less than $5.00 per gallon of fuel tank area to be protected. This means that a typical 4-place airplane with a 50 gallon fuel system would require **less than $250 in materials to protect it against post-crash**

explosion and fire. A HALON fire extinguishing system that provides HALON distribution nozzles for the passenger and baggage compartments, as well as for the engine compartment, costs less than $225. The effort to install these two systems is about 40 man hours per airplane.

That amounts to about the same price as six month's worth of insurance premiums! And the insurance premium dollars don't do a single thing toward stopping the actual post-crash fire and explosion.

The explosion-suppressing foam can easily be fitted in any fuel tank that is removable from the airplane. Flexible bladder tanks that are filled with foam, then encased in an aluminum outer shell are the best way to prevent post-crash explosions, but even existing wet-wing fuel tanks can be retrofitted with carved-to-fit foam blocks that can be inserted into the wing through the normal inspection plate openings in the wing tank. Wing airfoil templates can be made and placed on both ends of a rough-cut foam block, and the foam can be band-sawed or hot-wire cut to the fitted shape. Next, the air-foil shape can be sliced into smaller pieces that will fit through the inspection openings in the wing. (Imagine a loaf of bread that is sliced into 2-inch-thick or 4-inch-thick slices, then fitted into a cavity, in the same shape it was before it was cut into slices.)

It is important to provide cavities in the foam so the vent lines and drain lines will have about a teacup size void for fuel to vent and drain from, like a built-in sump. A similar cavity must be cut into the foam to give the fuel quantity float the necessary travel to measure fuel level. A piccolo tube type baffle must be placed around the fuel filler, so that incoming fuel and the filler

nozzle will not damage the foam in that area. All these cavities must be planned and trimmed in advance, of course, so the pieces will fit without the need to trim in place.

HALON 1211, 1301, AND HALOTRON EXTINGUISHING SYSTEMS

Current race car safety technology and rules dictate that all race cars in professional races **must** be equipped with on-board HALON fire extinguishing systems. This rule, plus the **mandatory** explosion-suppression foam-filled fuel tanks, have made fatal race car fires almost totally a thing of the past.

A particular HALON system that allows the pilot the option of momentary or full discharge of the HALON fire extinguishing system, is an option that is being incorporated into the airplane. This system places one discharge nozzle in the engine compartment where small fuel leak or oil leak fires could start, and two discharge nozzles in the passenger compartment where crash fires or hard-landing fires could be a threat to the pilot and passengers.

HALON is a gas that **partially** displaces oxygen in a fire, thereby depriving the fire of the oxygen ratio that supports combustion, while at the same time leaving just enough oxygen in the cabin atmosphere for human breathing and life-support.

This new, on-board fire-extinguishing system uses a HALON bottle mounted rigidly under the pilot's seat, with a toggle valve mounted on the instrument panel in plain view and easy reach of the pilot, and plumbing that directs the HALON to selected areas of the airplane. It puts out small fires and conserves HALON for future flare-ups, by just pressing the toggle valve and releasing it. It also has a lock and full-discharge options for large fires, allowing the pilot and passenger(s) time to escape. The HALON bottle can be partially or completely refilled at any fire extinguisher shop.

The HALON bottles come in sizes of 2.5 lb., 5.0 lb., 13.0 lb., and 20.0 lb. The retail cost ranges from $150 to $373, depending on the size of the bottle of HALON, and whether you add the extra toggle valve.

DUAL HALON 1211, 1301, HALOTRON CONTROLS

For about $50 extra, you can add a second toggle switch in your HALON system. One switch can control the engine fire extinguishing system separately from the cockpit of the airplane. Two toggle switches can conserve HALON, direct it to the known source of the fire, and can prevent HALON from entering the cabin if only an engine fire is in progress.

For those of you who are worried about the cost of the foam and HALON fire prevention systems...compare the cost to the cost of burn ward time in a hospital...or compare it to the cost of a human life.

SOURCE OF HALON 1211, 1301, HALOTRON SYSTEMS

HALOTRON
(not Halon)
BUCKEYE FIRE EQUIPMENT CO.
110 Kings Road
Kings Mountain, NC 28086

<u>HALON 1211,1301</u>
TRUECHOICE, INC.
4180 Weaver Court
Hilliard, Ohio 43206
1-800-388-8783

SUPER FLITE
2149 Eash Pratt Blvd.
Elk Grove Village, IL 60007
1-708-364-0858

McMASTER CARR SUPPLY
P.O. Box 54960
Los Angeles, CA 90054-0960

The builder of this Glassair T.D. is from Salinas, California, where the land is very flat. He has tested his Blanton Ford V-6 engine for many hours.

The instrument panel in the Legend V-8 has the six basic flight instruments for IFR flight, and at least 10 smaller instruments to display engine and system functions. A video camera would really be helpful to record the numbers when flight testing this airplane.

George Morse has tested this Rodeck V-8 small block powered Jaguar for hundreds of hours without any serious problems.

VIDEO RECORD

As noted at the beginning of this chapter, a video tape record of each of your flight tests will be very valuable in reviewing the progress of your flights. In most cases, the FAA or CAA in other countries, will require 35 hours of flight test time in a specific area near your home base airport.

Here in California, the FAA likes for us to stay over the flat valleys and away from the mountains. That is for two reasons. One, if your engine quits, a flat valley is a better place to have a forced landing than in the mountains. Second, if you crash, it is easier to find you if you are not in a mountainous area. This same reasoning prevails even if you are flying one of those certified airplane engines.

A video camera can be mounted on the co-pilot seat, the roll bar if you have one, or to a three point bracket that you can make especially for the video camera. Most video cameras mount at the bottom by a single 1/4" coarse thread bolt. Mount the camera so it shows the entire instrument panel and your hands if that is possible. You also want to verify that you can turn it on easily just before your first take-off roll starts. And mount it so that it will not vibrate. A roll of silver tape works great for quick vibration dampening.

147

The Dan Denney Thunder Mustang instrument panel has 4 IFR flight instruments, 7 engine and system instruments, plus a "glass panel" with a monitor screen that can present almost any function imaginable. It would still be a good idea to video tape your test flights, even with a panel like this.

Tracy Crook is shown flight testing his RV-4 with his Mazda rotary engine. Read his conversion book for an exciting description of one of his nearly disastrous test flights that ended safely and happily.
Tracy Crook Photo

MORE VIDEO

Another very important thing to do for your video flight test recording, is to assure that your camera battery will last 2 hours. If not, you need to get a 12 volt adapter to power your video camera off your aircraft battery.

Before you make your first flight you will surely do some high-speed taxi tests. Check out your video camera by taping those high speed taxi tests. If the camera does not work well or if it vibrates too much, you will know what to do to fix it before your first flight.

TAXI TESTS

In reading accident reports, you will always read about minor to serious accidents that happened while the pilot or A&P mechanic was merely doing a taxi test. Don't let the airplane lift off the runway until your first flight. Know the lift-off speed and keep the speed at least 10% to 15% below that for taxi tests. If your airplane lift-off speed is 60 mph, then don't get it over 50 mph during the taxi tests. It would be sad to wreck your airplane before you ever get it in the air.

EAA FLIGHT ADVISOR

Before you make your first flight arrange for a volunteer EAA Flight Advisor to check with you regarding your first flight plan of operation. You will be much safer if you do.

Companies & Suppliers

◆◆◆◆◆◆◆◆◆◆◆◆◆◆◆◆◆◆◆◆◆◆◆◆◆◆◆◆◆◆◆◆◆◆◆◆

A very good place to see the newest products for auto engine conversions such as this Stewart S-51 engine, gearbox and prop, would be at some of the larger fly-ins and air shows. Some of these shows are: Oshkosh, Wisconsin; Lakeland/Sun-N-Fun, Florida; Copperstate/Mesa, Arizona; Arlington, Washington; the Paris Air Show, Paris, France; and many other regional shows.

A LIVING LIST

The companies listed in this chapter were alive and well on the day the 4th edition went to the book bindery. But, this division of the aviation business is very dynamic.

Do not expect to see this list as 100% accurate in 5 years.

In looking back at the first edition of this book that was published in 1985, it is interesting to note that only 10% of the companies in business then, are still in business today.

The best way to know who is a viable business is to attend E.A.A. fly-ins and Air Shows. If you see the same company at the show this year, last year, and the year before, they are relatively stable.

Shirl Dickey sells plans and parts to build this Chevrolet V-6 powered E-RACER.

E.RACER - SHERIL DICKEY
Chevy V-6, Buick V-8 Engines
8631 W. College Dr.
Phoenix, AZ 85037
1-602-691-0515

SCOGGIN - DICKEY BUICK
G.M. ENGINES
5901 Spur 327
Lubbock, TX 79424
1-800-456-0211

ALTURAIR BD-5
350 hp Rotary Engines
1405 N. Johnson Ave
El Cahon, CA 92030
1-619-449-1570

BELTED AIR POWER
V-6 Chevy Conversions
1408 Western Ave.
Las Vegas, NV 89102
1-702-384-8006

CAM - FIRE GROUP
Honda Civic Conversions
9755 Saanich Rd.
Sidney, BC V8L-5T5 Canada
1-250-656-4774

ENGINEAIR, INC.
Chevrolet V-8 Conversions
209 Cessna Blvd.
Daytona Feach, FL 32124
1-904-788-2852

JONES CUSTOM AIRCRAFT
 SERVICE
Chevrolet V-8 Conversions
4010 S.W. 4th St.
Kissimmee, FL 34742
1-407-846-1224

AVTEC OF MONTANA
Honda Conversions
201 Cherry St. P.O.Box 138
Clancy, MT 59634-0138

ATKINS AVIATION, INC.
MAZDA PARU & Engines
16715 Meridian Ave. East, Bldg K
Puyallup, WA 98373
1-206-848-7776

EGGENFELLNER ADVANCED
 AIRCRAFT
Subaru Conversions
427-3 Amhurst St. Ste. 230
Nashua, NH 03063-1258
1-800-840-4620

ROSS AERO
Planetary Gear PSRU
3824 E. 37th. St.
Tucson, AZ 85713
1-520-747-7877

RFI
Subaru Conversions
P.O. Box 1444
McAlestar, OK 74502
1-918-823-4610

Fred Geschwender
Lincoln, Nebraska
Contact:
Jeff Ackland
Performance Aircraft
12901 W. 151st. St.
Suite C
Olathe, Kansas 66062
1-913-780-9140

SWAG Aeromotive
Electronic Fuel Injection
2521 North Fairview
Tucson, AZ 85705
1520-622-6910

Here you see a high-tech air cooled Volkswagen Beetle engine in Al Harrolson's Super Pup homebuilt airplane. With the help of Steve Parkman of SWAG Aeromotive of Tucson, AZ, Al installed a modern electronic fuel injection and ignition system on this airplane/auto engine conversion. Contact Steve at the address listed on this page to convert your auto, or even "certify" Lycoming to electronic fuel injection!

Steve Parkman, of SWAG Aeromotive, sells plans and books for converting GEO and SATURN engines.

George Morse (in white shirt, arms folded), sells complete kits to build his V-8 powered Jaguar airplane.

Merlin Aircraft sells kits to build this EXPLORER aircraft.

This Merlin Aircraft tail dragger airplane is powered by a MAZDA rotary engine. The MAZDA has a cog belt PSRU, designed by Richard Finch of FINCH ENGINEERING, Santa Barbara, CA. For plans to convert MAZDA 13B rotary engines for belt drive aircraft use, contact FINCH ENGINEERING.

FACT: At 12,500' altitude, your engine is producing 50% power.

LAPSE RATE: You lose 3.6% horsepower per 1,000 ft. gain in altitude.

FORMULA FOR HORSEPOWER

$$HP = \frac{TORQUE \times RPM}{5250}$$

BSFC: BRAKE SPECIFIC FUEL CONSUMPTION

$$= \frac{FUEL\ FLOW}{HP}$$

$$BMEP = \frac{150.8 \times TORQUE}{DISPLACEMENT}$$

Reiner Hoffman sells complete Subaru engines like this one in an Avid Flyer.

Tracy Crook sells books telling how to convert MAZDA rotary engines like this one.
Tracy Crook Photo

The Next Ten Years!

◆◆

The shape of General Aviation Manufacturers' Association for A.D. 2000 and beyond: Special Custom MT 4 blade prop, Orenda turbocharged, CERTIFIED V-8, pressurized, 2 place Lancair Tigress airframe. 400 mph? San Francisco to Denver in 3 hours, probably.

PREDICTING THE FUTURE

As you well know, if any of us could accurately predict the future, we could likely get rich! We could play the stock market and make fantastic deals. We could place bets on horses and win every time. These things are very unlikely.

But predicting the trend of a technology such as Converting Auto Engines for Experimental Aircraft is a lot easier to do than to accurately predict the winning horse in the Kentucky Derby five or ten years from now. We can all see where this auto engine technology is headed. So, here are some of my personal predictions for the future of auto engines in aircraft, for the next five to ten years. Ten years from now it will be fun to re-read these predictions.

FIREWALL FORWARD KITS

Take a tip from the pioneers in the auto engine conversion field, and believe that it takes a lot of time and money to design and flight test a truly dependable auto engine for a specific size and speed airplane. Ask Jess Meyers and Tom Jones of Belted Air Power about the trials and tribulations they experienced during the first 15 years of designing and flying their Buick and Chevrolet engines. They will tell you that it would have been much easier to write checks and buy a firewall forward engine and prop drive package.

Other companies, such as Stratus, Inc., operated by Reiner Hoffman, have perfected Subaru EA-81 and Subaru Legacy engines to the point that you can write a check and buy an auto engine kit that will bolt right in to certain airplanes.

Another company, Engineair, of Daytona Beach, Florida, can sell you a Chevrolet V-8 engine kit for your Lancair IV-P airplane. And this trend is increasing. Check chapter 19 for names and addresses of companies that presently offer firewall forward auto engine conversion kits. Expect to see more and more companies that will offer bolt in kits for almost all of the popular kitplanes that are available.

PURPOSE DESIGNED PLANES

One solution to several of the design considerations related to installing an auto engine in an airplane is to simply design a new airplane that provides for the slight extra weight of a cast iron auto engine, that provides for water cooling rather than air cooling, and that will be as dependable as our new cars are today.

This trend has been slowly growing even for as long as 60 years when Bernie Pietenpol designed his airplane to use Ford Model T and Ford Model A cast iron engines. Recently, Lancair has

This was a history making meeting on March 22, 1997. The EAA and the CAFE Foundation called this meeting to explore the future of auto engines for air races such as Reno, Phoenix, and more. Pictured here (left to right) are: Ryan Falconer, Falconer V-12; Dan Denney, Thunder Mustang; Scott Sedgewick, Orenda Engines; Darrel Buehl, Magnum Aircraft Engines; Ack Miller (standing), father of turbocharging; and Kevin Mc Clelland, of GM. Richard Finch was also a presenter. The verdict: auto engines can really fly in the future!

This Ryan Falconer V-12 in Dan Denney's Thunder Mustang will also make its way into certified airplanes such as Beechcraft Bonanzas, Piper Malibus, and even twin engine airplanes such as the Ted Smith Aerostar. Expect to see this happen.

The Orenda V-8 engine is actually being certified for use in King Air Twins to replace the expensive turboprops. Look for other auto engines to be certified soon.

designed a fast, 2-place pressurized plane expressly to use an Orenda V-8 water cooled engine. The author of this book has also designed a 4 place low wing, high speed airplane to specifically use a V-6 or V-8 water cooled auto engine. Expect to see a significant number of similar purpose-built airplanes to make their appearance in the next 10 years from the publication date of this book.

PURPOSE BUILT ENGINES

Expect to see a significant number of different custom built, custom assembled auto derived engines to be offered for sale. Already, the Falconer V-12 engine is astounding air show attendees with its sound and performance. The author of this book is offering blue-printed, balanced, and custom assembled Buick and Chevrolet V-6 engines for aircraft use. Jeff

Ackland offers Donovan/Brodix big block Chevrolet based engines for sale to the public for use in airplanes that need 450 to 650 horsepower. Custom built, blueprinted and balanced engines are really nice to fly because they are so powerful and so smooth.

TURBOCHARGING

Let's face it. Turbochargers were invented strictly for aircraft use.

They add significant power increases under all flight conditions, and they really make it better to fly out of high altitude airports and on hot days. Turbochargers can give you 100% power to 150% or more power and then you can always cruise a turbocharged aircraft engine at 50% to 75% power if you want to. The big thing that will make turbocharging more useful in the future is the new development in intercooling. When you cool the pressurized inlet air down to almost ambient temperature, you increase the engine power a lot with very little dependability problem. Expect to see water-alcohol injection, too. Expect to see more pressurized airplanes because of turbocharging.

ELECTRONICS

Face it folks. The electronic age arrived several years ago. Carburetors and ignition points and condensers are on the way out. As of 1990, you could not buy a new car in America that did not have electronic ignition and electronic fuel injection. Fear of lightning strikes can and will be overcome by proper shielding of the components. Lap top and palm held computers will be normal in our airplanes.

PROPELLERS

The number one dark ages item in a piston airplane is the prop. Expect to see dramatic innovations in airplane props. We sure do need new prop designs.

Look for dozens of Thunder Mustangs to be flying, even in special classes of all-Thunder Mustangs, at the Reno Air Races within 10 years.

Looking back 20 years or more, this Ford Escort conversion looks very crude. But it helped pave the way for more public acceptance of auto engines in airplanes today.

CERTIFIED AIRCRAFT ENGINES

For sure, some enterprising company will certify water cooled V-6 and V-8 piston engines for use in Pipers, Mooneys, Cessnas and Grummans. Certification can be done for $50,000 to $100,000, and it WILL be done!

Look for Supplemental type certificates that allow auto engines like this turbocharged, gear drive, fully muffled Corvair engine to be installed in Cessna 150 airplanes like this one, and flown as fully certified airplanes, including the auto engine. This conversion was the Bud Rinker gearbox drive Corvair engine, installed in the Cessna 150 by Richard Finch in 1980. The engine now powers an air boat in Florida, and the Cessna 150 is in Houston, Texas.